UNITEXT for Physics

UNITEXT for Physics series publishes textbooks in physics and astronomy, characterized by a didactic style and comprehensiveness. The books are addressed to upper-undergraduate and graduate students, but also to scientists and researchers as important resources for their education, knowledge, and teaching.

More information about this series at https://link.springer.com/bookseries/13351

Luca Salasnich

Modern Physics

Introduction to Statistical Mechanics,
Relativity, and Quantum Physics

 Springer

Luca Salasnich
Department of Physics and Astronomy
University of Padua
Padova, Italy

ISSN 2198-7882   ISSN 2198-7890   (electronic)
UNITEXT for Physics
ISBN 978-3-030-93745-4   ISBN 978-3-030-93743-0   (eBook)
https://doi.org/10.1007/978-3-030-93743-0

This Springer imprint is published by the registered company Springer Nature Switzerland AG
The registered company address is: Gewerbestrasse 11, 6330 Cham, Switzerland

Preface

This book contains the lecture notes prepared for two one-semester courses at the University of Padua: "Structure of Matter", B.Sc. in Optics and Optometrics, and "Quantum Physics", B.Sc. in Materials Science. These courses give an introduction to statistical mechanics, special and general relativity, and quantum physics.

Chapter 1 briefly reviews the ideas of classical statistical mechanics introduced by James Clerk Maxwell, Ludwig Boltzmann, Willard Gibbs, and others. Chapter 2 is mainly devoted to the special relativity of Albert Einstein but we briefly consider also the general relativity. In Chap. 3, the quantization of light due to Max Planck and Albert Einstein is historically analyzed, while Chap. 4 discusses the Niels Bohr quantization of the energy levels and the electromagnetic transitions. Chapter 5 investigates the Schrödinger equation, which was obtained by Erwin Schrödinger from the idea of Louis De Broglie to associate with each particle a quantum wavelength. Chapter 6 describes the basic axioms of quantum mechanics, which were formulated in the seminal books of Paul Dirac and John von Neumann. In this chapter, we also discuss the stationary perturbation theory, the time-dependent perturbation theory, and the variational principle. In Chap. 7, there are several important application of quantum mechanics: the quantum particle in a box, the quantum particle in the harmonic potential, and the quantum tunneling. Chapter 8 is devoted to the study of quantum atomic physics with special emphasis on the spin of the electron, which needs the Dirac equation for a rigorous theoretical justification. In Chap. 9, the quantum mechanics of many identical particles at zero temperature is explained, while in Chap. 10 the discussion is extended at finite temperature by introducing and using the quantum statistical mechanics. The appendices on Dirac delta function, complex numbers, Fourier transform, and differential equations are a useful mathematical aid for the reader.

The author acknowledges Dr. Fabio Sattin, Dr. Andrea Tononi, and Prof. Flavio Toigo for their critical reading of the manuscript and their useful comments and suggestions.

Padova, Italy
November 2021

Luca Salasnich

Contents

Chapter 1
Classical Statistical Mechanics

In this chapter we first discuss the kinetic theory of ideal gases and the Maxwell distribution of velocities at thermal equilibrium. Then we consider the more general Maxwell-Boltzmann distribution of non-interacting particles under the effect of an external trapping potential. Finally, we analyze the statistical ensembles of Gibbs, which are useful tools to connect the microscopic dynamics of interacting particles to the macroscopic behavior of a thermodynamical system.

1.1 Kinetic Theory of Gases

The kinetic theory of gases was formulated in the period between 1738 and 1871 with the contribution of several scientists, among them Daniel Bernoulli, Mikhail Lomonosov, August Krönig, Rudolf Clausius, James Clerk Maxwell, and Ludwig Boltzmann. This theory is the first historical example of statistical mechanics, where the macroscopic thermodynamics is described in terms of many microscopic particles (atoms or molecules). Actually, it was the description of the stochastic Brownian motion of a mesoscopic particle in a liquid, as due to the collisions with the microscopic particles of the liquid (made by Albert Einstein in 1905), that provided compelling proof that atoms and molecules exist. Jean Perrin confirmed this fact experimentally in 1908. In 1926, Perrin received the Nobel Prize in Physics "for his work on the discontinuous structure of matter".

At thermal equilibrium a very dilute gas is well described by the equation of state

$$P V = n R T , \tag{1.1}$$

where P is the pressure of the gas, V is the volume of the gas container, n is the number of moles, $R = 8.314 \, \text{J}/(\text{mol} \times \text{K})$ is the gas constant, and T is the absolute temperature (i.e. the temperature, usually measured in Kelvin, that is zero at the absolute zero, where the pressure of the ideal gas becomes zero). Equation (1.1) is known as the equation of state of ideal gases and it was formulated by Benoit Clapeyron in 1834.

August Krönig in 1856 and Rudolf Clausius in 1857 found, independently, that Eq. (1.1) can be derived from a microscopic kinetic theory. First of all, one observes that the number n of moles is related to the total number N of identical particles by the formula

$$n = \frac{N}{N_A},$$ (1.2)

where $N_A = 6.02 \cdot 10^{23}$ is the Avogadro number. After introducing the Boltzmann constant

$$k_B = \frac{R}{N_A} = 1.38 \cdot 10^{23} \, \text{J/K},$$ (1.3)

Equation (1.1) can be rewritten as

$$P V = N k_B T.$$ (1.4)

This equation clearly shows that the pressure P is proportional to the total number N of identical particles and to the absolute temperature T. Thus, it is quite natural to think that the pressure P exerted by the gas is due to the collisions of the particles on the container walls.

Let us now consider a cubic container of side L and volume $V = L^3$ with N identical particles of mass m inside. Let us choose the reference system with the Cartesian axes (x, y, x) along the sides of the box. The force $F_{i,x}$ that the i-th particle exerts, along the x direction on the container wall that parallel to the plane (y, z), is given by

$$F_{i,x} = \frac{\Delta(m v_{i,x})}{\Delta t} = \frac{2 m v_{i,x}}{\Delta t},$$ (1.5)

where $\Delta(m v_{i,x})$ is the variation of the linear momentum in the elastic collision of the i-th particle with the wall and Δt is the time interval. This time interval is not arbitrary if the particles are only interacting with the walls of the container. In this case

$$\Delta t = \frac{2L}{v_{i,x}},$$ (1.6)

that is the time interval between two collisions of the ith particle with the same wall. It then follows that

$$F_{i,x} = \frac{m v_{i,x}^2}{L}$$ (1.7)

and the pressure reads

$$P = \frac{\sum_{i=1}^{N} F_{i,x}}{L^2} = \frac{m}{L^3} \sum_{i=1}^{N} v_{i,x}^2 = \frac{m}{L^3} N \langle v_x^2 \rangle , \tag{1.8}$$

introducing the statistical average of a generic quantity A shared by the N identical particles as

$$\langle A \rangle = \frac{1}{N} \sum_{i=1}^{N} A_i . \tag{1.9}$$

Moreover, we assume independence with respect to the direction of propagation of the squared velocity, namely

$$\langle v^2 \rangle = \langle v_x^2 \rangle + \langle v_y^2 \rangle + \langle v_z^2 \rangle = 3 \langle v_x^2 \rangle . \tag{1.10}$$

It follows that the pressure P of Eq. (1.8) can be written as

$$P = \frac{mN}{3V} \langle v^2 \rangle . \tag{1.11}$$

Comparing Eq. (1.4) with Eq. (1.11) we obtain

$$\frac{1}{2} m \langle v^2 \rangle = \frac{3}{2} k_B T . \tag{1.12}$$

This remarkable formula relates the statistical average of the kinetic energy of the miscroscopic identical particles to the macroscopic absolute temperature T of the gas.

In this treatment the gas is indeed ideal because its total internal energy E is simply the sum of the kinetic energies $(1/2)mv_i^2$ of the single particles, i.e.

$$E = \sum_{i=1}^{N} \frac{1}{2} m v_i^2 = N \frac{1}{2} m \langle v^2 \rangle = N \frac{3}{2} k_B T = \frac{3}{2} n R T . \tag{1.13}$$

This is the correct formula for the internal energy of a monoatomic gas, where each atom has only three traslational degrees of freedom. In this case the *equipartition theorem* holds: at thermal equilibrium there is an associated thermal energy $k_B T / 2$ for each degree of freedom.

1.1.1 Maxwell Distribution of Velocities

In 1860 James Clerk Maxwell considered the probability distribution $f(\mathbf{v})$ of finding a particle with velocity $\mathbf{v}$ in a volume $d^3\mathbf{v}$ for the ideal gas at thermal equilibrium. Because $f(\mathbf{v})$ is a probability distribution it must satisfy the condition of normalization to one, namely

$$\int_{\mathbb{R}^3} f(\mathbf{v}) \, d^3\mathbf{v} = 1 \,, \tag{1.14}$$

where $\mathbb{R}^3$ is the three-dimensional space of velocities. Moreover, the statistical average of a generic observable $A(\mathbf{v})$, which is a function of $\mathbf{v}$, is defined as

$$\langle A(\mathbf{v}) \rangle = \int_{\mathbb{R}^3} A(\mathbf{v}) \, f(\mathbf{v}) \, d^3\mathbf{v} \,. \tag{1.15}$$

In particular, it follows that the statistical average of the square velocity v^2 reads

$$\langle v^2 \rangle = \int_{\mathbb{R}^3} v^2 \, f(\mathbf{v}) \, d^3\mathbf{v} \,. \tag{1.16}$$

Taking into account Eqs. (1.12) and (1.16), it follows that $f(\mathbf{v})$ must satisfy the crucial condition

$$\int_{\mathbb{R}^3} v^2 \, f(\mathbf{v}) \, d^3\mathbf{v} = 3 \frac{k_B T}{m} \,. \tag{1.17}$$

Each particle of the gas is characterized by its kinetic energy

$$\frac{1}{2} m v^2 = \frac{1}{2} m \left(v_x^2 + v_y^2 + v_z^2 \right) \,, \tag{1.18}$$

and, due to the isotropy of the problem with respect to the velocity, it is quite natural to assume that

$$f(\mathbf{v}) = C \, f_0(v^2) = C \, f_0(v_x^2 + v_y^2 + v_z^2) = C \, f_0(v_x^2) \, f_0(v_y^2) \, f_0(v_z^2) \,, \tag{1.19}$$

where C is a constant fixed by the normalization to one, Eq. (1.14). The only function $f_0(x)$ that satisfies the equation

$$f_0(x + y + z) = f_0(x) \, f_0(y) \, f_0(z) \tag{1.20}$$

is the exponential function, i.e.

$$f_0(x) = e^{\alpha x} \,, \tag{1.21}$$

and it means that

$$f_0(v^2) = e^{\alpha \left(v_x^2 + v_y^2 + v_z^2 \right)} \,, \tag{1.22}$$

where α is a constant fixed by Eq. (1.17). It is then straightforward to find

$$C = \left(\frac{m}{2\pi k_B T} \right)^{3/2} \qquad \text{and} \qquad \alpha = -\frac{m}{2 k_B T} \,. \tag{1.23}$$

Fig. 1.1 Maxwell
distribution $\Phi(v)$ of the
particle speed $v = |\mathbf{v}|$ for
three values of the
temperature T. In the plot we
choose units such that $m-1$

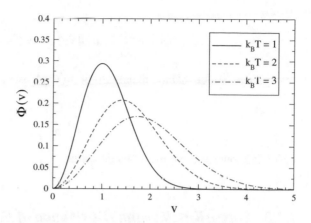

In conclusion, the Maxwell distribution of the velocities is given by

$$f(\mathbf{v}) = \left(\frac{m}{2\pi k_B T}\right)^{3/2} e^{-\frac{mv^2}{2k_B T}} . \tag{1.24}$$

For historical reasons one usually introduces the parameter

$$\beta = \frac{1}{k_B T} \tag{1.25}$$

and the Maxwell distribution then reads

$$f(\mathbf{v}) = \left(\frac{m\beta}{2\pi}\right)^{3/2} e^{-\beta\frac{mv^2}{2}} . \tag{1.26}$$

It is important to stress that, adopting spherical coordinates and taking into account the spherical symmetry of the problem we have $d^3\mathbf{v} = 4\pi v^2 dv$, and we can also introduce

$$\Phi(v) = 4\pi v^2 f(|\mathbf{v}|) = 4\pi \left(\frac{m\beta}{2\pi}\right)^{3/2} v^2 e^{-\beta\frac{mv^2}{2}} \tag{1.27}$$

that is the probability distribution of the modulus $v = |\mathbf{v}|$ of the velocity $\mathbf{v}$, and it is such that (Fig. 1.1)

$$\int_0^{+\infty} \Phi(v)\, dv = 1 . \tag{1.28}$$

We immediately find

$$\langle v \rangle = \int_0^{+\infty} v\, \Phi(v)\, dv = \sqrt{\frac{8}{\pi}} \sqrt{\frac{k_B T}{m}} \tag{1.29}$$

and

$$\langle v^2 \rangle = \int_0^{+\infty} v^2\, \Phi(v)\, dv = 3\frac{k_B T}{m}\,, \tag{1.30}$$

as expected. Notice that the maximum of $\Phi(v)$ is instead obtained at

$$v_{mp} = \sqrt{\frac{2k_B T}{m}}\,, \tag{1.31}$$

which is known as the most probable speed.

1.1.2 Maxwell-Boltzmann Distribution of Energies

In 1872 Ludwig Boltzmann analyzed an ideal gas of identical particles of mass m under the action of an external potential energy $U(\mathbf{r})$, for instance the gravitational potential energy $U(\mathbf{r}) = mgz$ with gravity acceleration $g = 9.81$ m/s^2. In this case the total internal energy of the gas is given by

$$E = \sum_{i=1}^{N} \varepsilon(\mathbf{r}_i, \mathbf{p}_i)\,, \tag{1.32}$$

where

$$\varepsilon(\mathbf{r}, \mathbf{p}) = \frac{p^2}{2m} + U(\mathbf{r}) \tag{1.33}$$

is the single-particle energy with $\mathbf{p} = m\mathbf{v}$ the linear momentum of the generic particle and $\mathbf{r}$ its position vector. Notice that $p^2/(2m) = mv^2/2$.

Boltzmann introduced the adimensional probability distribution $f(\mathbf{r}, \mathbf{v})$ of finding a particle with position $\mathbf{r}$ and linear momentum $\mathbf{p}$ in this many-particle system at thermal equilibrium. Because $f(\mathbf{r}, \mathbf{p})$ is a probability distribution it must satisfy the condition of normalization to one, namely

$$\int_{\mathcal{V}} f(\mathbf{r}, \mathbf{v})\, \frac{d^3\mathbf{r}\, d^3\mathbf{p}}{h^3} = 1\,, \tag{1.34}$$

where $\mathcal{V}$ is the so-called single-particle phase space volume and h is a constant introduced to make the infinitesimal element $d^3\mathbf{r}d^3\mathbf{p}/h^3$ adimensional. The constant h, which must be an action, i.e. a quantity with units Joule $\times$ seconds, will be identified as the Planck constant ($h = 6.63 \cdot 10^{-34}$ J$\times$s) within the quantum statistical mechanics. In full generality $\mathcal{V} = \mathbb{R}^3 \times \mathbb{R}^3$ but if the particles are confined in a volume V one has $\mathcal{V} = V \times \mathbb{R}^3$. In the absence of the external potential one must recover the Maxwell distribution of Eq. (1.26) and, consequently, Boltzmann suggested the following more general Maxwell-Boltzmann distribution

$$f(\mathbf{r}, \mathbf{p}) = \frac{1}{\mathcal{Z}_1} e^{-\beta \varepsilon(\mathbf{r}, \mathbf{p})} , \tag{1.35}$$

where $\mathcal{Z}_1$, that is called single-particle partition function, is determined by the normalization (1.34), i.e.

$$\mathcal{Z}_1 = \int_V e^{-\beta \varepsilon(\mathbf{r}, \mathbf{p})} \frac{d^3 \mathbf{r} \, d^3 \mathbf{p}}{h^3} . \tag{1.36}$$

According to Boltzmann, for a gas of non-interacting particles, the statistical average of a generic observable $A(\mathbf{r}, \mathbf{p})$, which is a function of $\mathbf{r}$ and $\mathbf{p}$, is defined as

$$\langle A(\mathbf{r}, \mathbf{p}) \rangle = \int_V A(\mathbf{r}, \mathbf{p}) \, f(\mathbf{v}) \frac{d^3 \mathbf{r} \, d^3 \mathbf{p}}{h^3} , \tag{1.37}$$

namely

$$\langle A(\mathbf{r}, \mathbf{p}) \rangle = \frac{1}{\mathcal{Z}_1} \int_V A(\mathbf{r}, \mathbf{p}) \, e^{-\beta \varepsilon(\mathbf{r}, \mathbf{p})} \frac{d^3 \mathbf{r} \, d^3 \mathbf{p}}{h^3} \tag{1.38}$$

Many experiments have shown that the Maxwell-Boltzmann distribution (1.35) is extemely accurate for non-interacting identical particles. However, at very low temperature the Maxwell-Boltzmann distribution is not reliable because of quantum mechanical effects, which are the main topic of this book.

1.1.3 Single-Particle Density of States

In many applications the single-particle partition function (1.36) is written in a compact way as

$$\mathcal{Z}_1 = \int_0^{+\infty} \mathcal{D}_1(\varepsilon) \, e^{-\beta \varepsilon} \, d\varepsilon , \tag{1.39}$$

where

$$\mathcal{D}_1(\varepsilon) = \int_V \delta(\varepsilon - \varepsilon(\mathbf{r}, \mathbf{p})) \frac{d^3 \mathbf{r} \, d^3 \mathbf{p}}{h^3} \tag{1.40}$$

is the so-called single-particle density of states with $\delta(x)$ the Dirac delta function. Actually, in this way the expression of $\mathcal{Z}_1$ seems simple but, in general, the calculation of the density of states $\mathcal{D}_1(\varepsilon)$ is not.

As an example, let us suppose that $\varepsilon(\mathbf{r}, \mathbf{p}) = p^2/(2m)$ and $\mathcal{V} = V \times \mathbb{R}^3$. Then we have

$$\mathcal{D}_1(\varepsilon) = \int_{V \times \mathbb{R}^3} \delta\left(\varepsilon - \frac{p^2}{2m}\right) \frac{d^3\mathbf{r}\, d^3\mathbf{p}}{h^3} = \frac{4\pi V}{h^3} \int_0^{+\infty} p^2\, \delta\left(\varepsilon - \frac{p^2}{2m}\right) dp$$

$$= \frac{2\pi V}{h^3} (2m)^{3/2} \int_0^{+\infty} y^{1/2}\, \delta(\varepsilon - y)\, dy = \frac{2\pi V}{h^3} (2m)^{3/2}\, \varepsilon^{1/2}. \qquad (1.41)$$

This is the single-particle density of states of a gas of ideal identical particles of mass m moving inside a box of volume V.

1.2   Statistical Ensembles of Gibbs

The total energy of Eq. (1.32) is separable in the contribution of single-particle energies because the particles are not interacting each other. In full generality one must take into account the interaction potential $V(\mathbf{r}_i, \mathbf{r}_j)$ between particles and the total energy of the system of N identical interacting particles of mass m reads

$$E = \sum_{i=1}^{N} \left[\frac{p_i^2}{2m} + U(\mathbf{r}_i)\right] + \frac{1}{2} \sum_{\substack{i,j=1 \\ i \neq j}}^{N} V(\mathbf{r}_i, \mathbf{r}_j) = H(\vec{\mathbf{r}}, \vec{\mathbf{p}}), \qquad (1.42)$$

where the total energy is a function of $\vec{\mathbf{r}} = (\mathbf{r}_1, \mathbf{r}_2, ..., \mathbf{r}_{N-1}, \mathbf{r}_N)$ and $\vec{\mathbf{p}} = (\mathbf{p}_1, \mathbf{p}_2, ..., \mathbf{p}_{N-1}, \mathbf{p}_N)$ which are vectors with 3N components. Here we have used the symbol $H(\vec{\mathbf{r}}, \vec{\mathbf{p}})$ to represent the functional dependence of the energy with respect to the $6N$ variables. When the energy is written in terms of coordinates and linear momenta it is said to be the Hamiltonian function H of the system.

In 1878 Joisiah Willard Gibbs introduced the concept of *statistical ensemble*: a large number of virtual copies of a macroscopic system, each of which represents a possible microscopic state that the macroscopic system might be in. Gibbs proposed different statistical ensembles and here we discuss the three most used.

1.2.1   Microcanonical Ensemble

The microcanonical ensemble describes an isolated system characterized by a fixed volume V, a fixed number N of identical particles, and a fixed total energy E. According to Gibbs, for a gas of interacting particles in the microcanonical ensemble, the statistical average of a generic observable $A(\vec{\mathbf{r}}, \vec{\mathbf{p}})$ is defined as

$$\langle A(\vec{\mathbf{r}}, \vec{\mathbf{p}}) \rangle = \frac{1}{\mathcal{W}} \int_{V^N \times \mathbb{R}^{3N}} A(\vec{\mathbf{r}}, \vec{\mathbf{p}})\, \mathcal{W}_0\, \delta\left(E - H(\vec{\mathbf{r}}, \vec{\mathbf{p}})\right) \frac{d^{3N}\vec{\mathbf{r}}\, d^{3N}\vec{\mathbf{p}}}{N!\, h^{3N}}, \qquad (1.43)$$

where $\delta(x)$ is the Dirac delta function, $\mathcal{W}_0$ is an arbitrary constant with the units of energy such that the statistical density function $\mathcal{W}_0\,\delta\,(E - H(\vec{r}, \vec{p}))$ is adimensional, and the key quantity is the number $\mathcal{W}$ of microscopic states (microstates) which correspond to the same macroscopic state:

$$\mathcal{W} = \int_{V^N \times \mathbb{R}^{3N}} \mathcal{W}_0\,\delta\,(E - H(\vec{r}, \vec{p}))\,\frac{d^{3N}\vec{r}\,d^{3N}\vec{p}}{N!\,h^{3N}}, \tag{1.44}$$

It is important to stress that the factorial term $N!$ which appears in Eqs. (1.43) and (1.44) is know as the correct counting term of Gibbs, and it takes into account that identical particles must be indistinguishable. Indeed, $N!$ is the number of permutations of N particles.

In the microcanonical ensemble of Gibbs the connection with the thermodynamics is simply given by

$$S = k_B \ln{(\mathcal{W})}, \tag{1.45}$$

which introduces the entropy S as a function of energy E, volume V and number N of particles. Note that this equation was discovered by Ludwig Boltzmann in 1872 and that the correct counting term $N!$ of Gibbs is really crucial to get an extensive entropy, i.e. an entropy whose value is proportional to the size of the system.

It is important to remember that Rudolf Clausius in 1850 made the first complete formulation of the *first law of thermodynamics*: the infinitesimal heat dQ absorbed by a macroscopic system produces an infinitesimal work dW done by the macroscopic system or an infinitesimal growth dE of its internal energy. In symbols,

$$dQ = dW + dE. \tag{1.46}$$

In addition, if this infinitesimal transformation is reversible, then

$$dQ = T\,dS, \tag{1.47}$$

where T is the temperature and dS is an infinitesimal variation of the entropy S. Moreover, in many cases the work can be written as

$$dW = P\,dV - \mu\,dN \tag{1.48}$$

where P is the pressure, dV is an infinitesimal variation of the volume, μ is the chemical potential, and dN is the corresponding variation of the number of particles. Under these conditions Eq. (1.46) gives

$$dS = \frac{1}{T}dE + \frac{P}{T}dV - \frac{\mu}{T}dN. \tag{1.49}$$

Now, from the entropy $S(E, V, N)$ as function of internal energy E, volume V, and number N of particles, we also have

$$dS = \left(\frac{\partial S}{\partial E}\right)_{V,N} dE + \left(\frac{\partial S}{\partial V}\right)_{E,N} dV + \left(\frac{\partial S}{\partial N}\right)_{E,V} dN. \tag{1.50}$$

Comparing Eq. (1.49) with Eq. (1.50) we obtain the absolute temperature T, the pressure P and the chemical potential μ as

$$\frac{1}{T} = \left(\frac{\partial S}{\partial E}\right)_{V,N}, \qquad P = T\left(\frac{\partial S}{\partial V}\right)_{E,N}, \qquad \mu = -T\left(\frac{\partial S}{\partial N}\right)_{E,V}, \tag{1.51}$$

which are familiar relationships of equilibrium thermodynamics. In the microcanonical ensemble the independent thermodynamic variables are E, N and V, while T, P and μ are dependent thermodynamic variables.

1.2.2 Canonical Ensemble

The canonical ensemble describes a system characterized by a fixed volume V, a fixed number N of identical particles, and a fixed temperature T due to the thermal contact with a heat bath. According to Gibbs, for a gas of particles in the canonical ensemble, the statistical average of a generic observable $A(\vec{r}, \vec{p})$ is defined as

$$\langle A(\vec{r}, \vec{p}) \rangle = \frac{1}{\mathcal{Z}_N} \int_{V^N \times \mathbb{R}^{3N}} A(\vec{r}, \vec{p})\, e^{-\beta H(\vec{r}, \vec{p})}\, \frac{d^{3N}\vec{r}\, d^{3N}\vec{p}}{N!\, h^{3N}}, \tag{1.52}$$

where $\mathcal{Z}_N$ is the canonical partition function defined as

$$\mathcal{Z}_N = \int_{V^N \times \mathbb{R}^{3N}} e^{-\beta H(\vec{r}, \vec{p})}\, \frac{d^{3N}\vec{r}\, d^{3N}\vec{p}}{N!\, h^{3N}}, \tag{1.53}$$

which is clearly a generalization of the single-particle partition function $\mathcal{Z}_1$ of Eq. (1.53).

In the canonical ensemble, the connection between statistical mechanics and thermodynamics is due to the formula

$$\mathcal{Z}_N = e^{-\beta F}, \tag{1.54}$$

which introduces the Helmholtz free energy F, whose thermodynamical definition is

$$F = E - TS, \tag{1.55}$$

where E is the internal energy, T is the temperature and S is the entropy. Here F is as a function of temperature T, volume V and number N of particles. From the Helmholtz free energy $F(T, V, N)$ the entropy S, the pressure P and the chemical potential μ are obtained as

$$S = -\left(\frac{\partial F}{\partial T}\right)_{V,N}, \quad P = -\left(\frac{\partial F}{\partial V}\right)_{T,N}, \quad \mu = \left(\frac{\partial F}{\partial N}\right)_{T,V}, \quad (1.56)$$

which are standard relationships of equilibrium thermodynamics such that

$$dF = -SdT - PdV + \mu dN, \quad (1.57)$$

that is another formulation of the first law of thermodynamics. In the canonical ensemble the independent thermodynamic variables are T, N and V, while S, P and μ are dependent thermodynamic variables.

1.2.3 Grand Canonical Ensemble

The grand canonical ensemble describes an open system characterized by a fixed volume V, a fixed temperature T, and a fixed chemical potential μ due to the weak thermal-chemical contact with a reservoir. According to Gibbs, for a gas of interacting particles in the grand canonical ensemble, the statistical average of a generic observable $A(\vec{r}, \vec{p})$ is defined as

$$\langle A(\vec{r}, \vec{p}) \rangle = \frac{1}{\mathcal{Z}} \sum_{N=0}^{+\infty} \int_{V^N \times \mathbb{R}^{3N}} A(\vec{r}, \vec{p}) \, e^{-\beta(H(\vec{r},\vec{p})-\mu N)} \frac{d^{3N}\vec{r} \, d^{3N}\vec{p}}{N! \, h^{3N}}, \quad (1.58)$$

where $\mathcal{Z}$ is the grand canonical partition function defined as

$$\mathcal{Z} = \sum_{N=0}^{+\infty} e^{\beta \mu N} \, \mathcal{Z}_N, \quad (1.59)$$

namely as

$$\mathcal{Z} = \sum_{N=0}^{+\infty} \int_{V^N \times \mathbb{R}^{3N}} e^{-\beta(H(\vec{r},\vec{p})-\mu N)} \frac{d^{3N}\vec{r} \, d^{3N}\vec{p}}{N! \, h^{3N}}. \quad (1.60)$$

In the grand canonical ensemble the connection between the statistical mechanics and the thermodynamics is due to the formula

$$\mathcal{Z} = e^{-\beta \Omega}, \quad (1.61)$$

which introduces the grand potential Ω, whose thermodynamical definition is

$$\Omega = E - TS - \mu N. \quad (1.62)$$

Here Ω is a function of temperature T, volume V and chemical potential μ. From the grand potential $\Omega(T, V, \mu)$ the entropy S, the pressure P and the number N of particles are obtained as

$$S = -\left(\frac{\partial \Omega}{\partial T}\right)_{V,\mu}, \qquad P = -\left(\frac{\partial \Omega}{\partial V}\right)_{T,\mu}, \qquad N = -\left(\frac{\partial \Omega}{\partial \mu}\right)_{V,T}, \qquad (1.63)$$

which are standard relationships of equilibrium thermodynamics such that

$$d\Omega = -S\, dT - P\, dV - N\, d\mu, \qquad (1.64)$$

that is an alternative formulation of the first law of thermodynamics. In the grand canonical ensemble the independent thermodynamic variables are T, μ and V, while S, P and N are dependent thermodynamic variables.

1.2.4 Many-Particle Density of States

Clearly, the canonical partition function (1.53) can be written as

$$\mathcal{Z}_N = \int_0^{+\infty} \mathcal{D}_N(\varepsilon)\, e^{-\beta\varepsilon}\, d\varepsilon, \qquad (1.65)$$

where

$$\mathcal{D}_N(\varepsilon) = \int_{V^N \times \mathbb{R}^{3N}} \delta(\varepsilon - H(\vec{\mathbf{r}}, \vec{\mathbf{p}}))\, \frac{d^{3N}\vec{\mathbf{r}}\, d^{3N}\vec{\mathbf{p}}}{N!\, h^{3N}}, \qquad (1.66)$$

is the many-particle density of states. Similarly, the grand canonical partition function (1.60) can be written as

$$\mathcal{Z} = \sum_{N=0}^{+\infty} \int_0^{+\infty} \mathcal{D}_N(\varepsilon)\, e^{-\beta(\varepsilon - \mu N)}\, d\varepsilon. \qquad (1.67)$$

1.3 Heat Capacity of Gases and Solids

The molar heat capacity at constant volume V is defined as

$$c_V = \frac{1}{n}\left(\frac{dQ}{dT}\right)_{V,N}, \qquad (1.68)$$

where dQ is the amount of exchanged heat and dT is the corresponding variation of temperature. Because $n = N/N_A$ with N_A the Avogadro number, and here $dV = dN = 0$, from Eq. (1.48) it follows $dQ = dE$, i.e.

$$c_V = \frac{1}{n}\left(\frac{dE}{dT}\right)_{V,N}.$$ (1.69)

For an ideal monoatomic gas the internal energy is given by $E = (3/2)nRT$ and consequently

$$c_V = \frac{3}{2}R.$$ (1.70)

This result is in very good agreement with the experimental data of monoatomic gases at room temperature, and also at high temperature. However, at low temperature one finds that this formula fails. The molar heat capacity of the monoatomic gas can also be deduced invoking the *equipartition theorem* of classical statistical mechanics saying that for each quadratic degree of freedom there is an associated thermal energy $k_B T/2$. Indeed, in the case of a free atom there are 3 translational degrees of freedom, which are 3 quadratic components of the velocity.

In a crystalline solid the atoms are distributed in a ordered and periodic structure. These atoms remain, on the average, in a specific site of the crystal lattice but they can oscillate around their equilibrium position. Thus, the motion of each atom in a solid is similar to the one of a harmonic oscillator, which is characterized by $3 + 3$ degrees of freedom: 3 for the kinetic energy and 3 for the elastic potential energy. Using again the equipartition theorem one immediately gets

$$c_V = 3R$$ (1.71)

because $E = 6N(k_B T/2)$ and $c_V = 3Nk_B/n = 3Nk_B/(N/N_A) = 3k_B N_A = 3R$. This expression for the molar heat capacity at constant volume of a solid is know as Dulong-Petit law. It was proposed in 1819 by Pierre Louis Dulong and Alexis Therese Petit as simple empirical fitting formula of their eperimental data at room temperature. Subsequent low-temperature studies showed that the Dulong-Petis law always fails, provided the temperature is low enough.

We have seen two examples of physical systems (monoatomic gas and crystalline solid) where the molar specific heat predicted by classical statistical mechanics is in disagreement with experiments at low temperature. In Chaps. 4 and 10 we will see that quantum statistical mechanics is the appropriate theory to explain these empirical evidences.

Further Reading

Two excellent books on statistical mechanics are:
Landau, L.D., Lifshitz, E.M.: Statistical Physics. Pergamon (1980).
Huang, K.: Statistical Mechanics. Wiley (1987).
Relevant historical papers are:
Maxwell, J.C.: Philos. Trans. R. Soc. Lond. **157**, 49 (1867).
Boltzmann, L.: Sitzungsberichte der Kaiserlichen Akademie der Wissenschaften in
Wien, Mathematisch-Naturwissenschaftliche Classe, **66**, 275 (1872).
The statistical ensembles were introduced by Gibbs in his book:
Gibbs, J.W.: Elementary Principles in Statistical Mechanics. Charles Scribner's Sons
(1902); Cambridge University Press (2010).

Chapter 2
Special and General Relativity

In this chapter we review the main results of the theory of special relativity, which was developed by Einstein in 1905. Special relativity has a wide range of consequences which have been confirmed by experiments. Among them, the universal speed limit, the length contraction, the time dilation, the mass-energy equivalence, and the formulas of relativistic dynamics. For the sake of completeness, in the last section we briefly discuss the theory of general relativity, that is the geometric theory of gravitation formulated by Einstein in 1916.

2.1 Electromagnetic Waves

In 1861 James Clerk Maxwell wrote a set of 20 coupled partial differential equations which are the foundation of classical electromagnetism. Some years later Oliver Heaviside restructured Maxwell original equations to be the 4 equations that we now recognize as Maxwell's equations. In the vacuum these equations are

$$\nabla \cdot \mathbf{E} = \frac{\rho}{\varepsilon_0} , \tag{2.1}$$

$$\nabla \cdot \mathbf{B} = 0 , \tag{2.2}$$

$$\nabla \wedge \mathbf{E} = -\frac{\partial \mathbf{B}}{\partial t} , \tag{2.3}$$

$$\nabla \wedge \mathbf{B} = \mu_0 \mathbf{j} + \varepsilon_0 \mu_0 \frac{\partial \mathbf{E}}{\partial t} , \tag{2.4}$$

where $\mathbf{E}(\mathbf{r}, t)$ is the electric field, $\mathbf{B}(\mathbf{r}, t)$ is the magnetic field, $\rho(\mathbf{r}, t)$ is the electric charge density, and $\mathbf{j}(\mathbf{r}, t)$ is the electric current density. ε_0 is the dielectric constant

© The Author(s), under exclusive license to Springer Nature Switzerland AG 2022
L. Salasnich, *Modern Physics*, UNITEXT for Physics,
https://doi.org/10.1007/978-3-030-93743-0_2

of vacuum, given by $\varepsilon_0 = 8.85 \cdot 10^{-12}$ C^2/(N×m^2). Instead, μ_0 is the paramagnetic constant of vacuum, given by $\mu_0 = 4\pi \cdot 10^{-7}$ V×s/(amp×m).

In the absence of electric sources ($\rho = 0$, $\mathbf{j} = \mathbf{0}$) Maxwell's equations are much simpler and read

$$\nabla \cdot \mathbf{E} = 0, \tag{2.5}$$

$$\nabla \cdot \mathbf{B} = 0, \tag{2.6}$$

$$\nabla \wedge \mathbf{E} = -\frac{\partial \mathbf{B}}{\partial t}, \tag{2.7}$$

$$\nabla \wedge \mathbf{B} = \varepsilon_0 \, \mu_0 \, \frac{\partial \mathbf{E}}{\partial t}. \tag{2.8}$$

Maxwell found that the electric and magnetic fields of Eqs. (2.5)–(2.8) satisfy the wave equations

$$\left(\frac{1}{c^2} \frac{\partial^2}{\partial t^2} - \nabla^2 \right) \mathbf{E} = \mathbf{0}, \tag{2.9}$$

$$\left(\frac{1}{c^2} \frac{\partial^2}{\partial t^2} - \nabla^2 \right) \mathbf{B} = \mathbf{0}, \tag{2.10}$$

where

$$c = \frac{1}{\sqrt{\varepsilon_0 \mu_0}} = 3 \cdot 10^8 \text{ m/s} \tag{2.11}$$

is the speed of light in vacuum. It is important to stress that, strictly speaking, since 1983 the meter is defined in the International System of Units as the distance light travels in vacuum in 1/299792458 of a second. This definition fixes the speed of light in vacuum at exactly 299792458 m/s. From Eqs. (2.9) to (2.10) Maxwell concluded that light is an electromagnetic field characterized by the coexisting presence of an electric field $\mathbf{E}(\mathbf{r}, t)$ and a magnetic field $\mathbf{B}(\mathbf{r}, t)$. Note that Eqs. (2.9) and (2.10) are usually called d'Alembert equations of electromagnetic waves.

As an exercise, we now derive Eq. (2.9) from Maxwell's equations. Inserting the curl into Eq. (2.7) we get

$$\nabla \wedge (\nabla \wedge \mathbf{E}) = \nabla \wedge \left(-\frac{\partial \mathbf{B}}{\partial t} \right) = -\frac{\partial}{\partial t} (\nabla \wedge \mathbf{B}). \tag{2.12}$$

For any vector field $\mathbf{E}(\mathbf{r}, t)$, an identity of differential calculus gives

$$\nabla \wedge (\nabla \wedge \mathbf{E}) = -\nabla^2 \mathbf{E} + \nabla (\nabla \cdot \mathbf{E}). \tag{2.13}$$

Therefore, by using Eq. (2.5), for the electric field we obtain

$$\nabla \wedge (\nabla \wedge \mathbf{E}) = -\nabla^2 \mathbf{E} \tag{2.14}$$

and Eq. (2.12) becomes

$$- \nabla^2 \mathbf{E} = - \frac{\partial}{\partial t} \left(\mathbf{\nabla} \wedge \mathbf{B} \right) . \tag{2.15}$$

Finally, taking into account Eq. (2.8), we can write

$$- \nabla^2 \mathbf{E} = - \varepsilon_0 \mu_0 \frac{\partial^2 \mathbf{E}}{\partial t^2} , \tag{2.16}$$

which is exactly Eq. (2.9). In a similar way one can easily derive also Eq. (2.10).

Equations (2.9) and (2.10), which are fully confirmed by experiments, admit monochromatic complex plane wave solutions

$$\mathbf{E}(\mathbf{r}, t) = \mathbf{E}_0 \, e^{i(\mathbf{k} \cdot \mathbf{r} - \omega t)} , \tag{2.17}$$

$$\mathbf{B}(\mathbf{r}, t) = \mathbf{B}_0 \, e^{i(\mathbf{k} \cdot \mathbf{r} - \omega t)} , \tag{2.18}$$

where $\mathbf{k}$ is the wavevector and ω the angular frequency, such that

$$\omega = c k , \tag{2.19}$$

is the dispersion relation, with $k = |\mathbf{k}|$ is the wavenumber. From Maxwell's equations one finds that the vectors $\mathbf{E}(\mathbf{r}, t)$ and $\mathbf{B}(\mathbf{r}, t)$ of these electromagnetic waves are mutually orthogonal. In addition they are transverse fields, i.e. orthogonal to the wavevector $\mathbf{k}$, which gives the direction of propagation of the wave. For completeness, let us remind that the wavelength λ is given by

$$\lambda = \frac{2\pi}{k} , \tag{2.20}$$

and that the linear frequency ν and the angular frequency $\omega = 2\pi\nu$ are related to the wavelength λ and to the wavenumber k by the formulas

$$\lambda \nu = \frac{\omega}{k} = c . \tag{2.21}$$

Assuming that the electromagnetic plane waves are confined in a cubic region of volume $V = L^3$ and imposing periodic boundary conditions, i.e. $\mathbf{E}(x + L, y + L, z + L, t) = \mathbf{E}(x, y, z, t)$ and $\mathbf{B}(x + L, y + L, z + L, t) = \mathbf{B}(x, y, z, t)$, we immediately find from Eqs. (2.17) and (2.18) that the wavevector $\mathbf{k}$ is quantized, namely

$$\mathbf{k} = \frac{2\pi}{L} \mathbf{n} , \tag{2.22}$$

where $\mathbf{n} = (n_1, n_2, n_3) \in \mathbb{Z}^3$. In this case, the sum with respect to all the wavevectors is usually written as

$$\sum_k = \sum_n = \sum_{n_1=-\infty}^{+\infty} \sum_{n_2=-\infty}^{+\infty} \sum_{n_3=-\infty}^{+\infty} . \qquad (2.23)$$

It follows that in the continuum limit one gets

$$\sum_k = \sum_n \to \int d^3\mathbf{n} = L^3 \int \frac{d^3\mathbf{k}}{(2\pi)^3} = V \int \frac{d^3\mathbf{k}}{(2\pi)^3} . \qquad (2.24)$$

2.1.1   Lorentz Invariance of d'Alembert Operator

In 1887 Albert Michelson and Edward Morley made a break-through experiment of optical interferometry showing that the speed of light in the vacuum is always given by Eq. (2.11) independently on the relative motion of the observer. Two years later, Henry Poincaré suggested that the speed of light is the maximum possible value for any kind of velocity. On the basis of previous ideas of George Francis FitzGerald, in 1904 Hendrik Lorentz found that Maxwell's equations of electromagnetism, and consequently also d'Alembert's equations (2.9) and (2.10), are not invariant with respect to the familiar Galilei transformations

$$x' = x - vt , \qquad (2.25)$$
$$y' = y , \qquad (2.26)$$
$$z' = z , \qquad (2.27)$$
$$t' = t . \qquad (2.28)$$

In the Galilei transformations, one considers a particle P measured by two orthogonal Cartesian reference systems O and O'. Within this framework, $OP = \mathbf{r} = (x, y, z)$ is the position vector and t is the time coordinate of the particle observed bt the reference system O. Instead, $O'P = \mathbf{r}' = (x', y', z')$ is the position vector and t' is the time coordinate of a particle observed by the reference system O'. Moreover, here we have assumed that the two reference systems are moving with relative velocity $\mathbf{v} = (v, 0, 0)$, such that at $t = t' = 0$ the two reference systems are coincident and during the time evolution they simply translate each other along the X axis.

Hendrik Lorentz considered the more general transformations

$$x' = \gamma(x - v t) , \qquad (2.29)$$
$$y' = y , \qquad (2.30)$$
$$z' = z , \qquad (2.31)$$
$$t' = \gamma \left(t - \beta \frac{x}{c} \right) , \qquad (2.32)$$

where γ and β are unknown dimensionless parameters. Following Lorentz, we can ask ourself what are the values of γ and β for which the d'Alembert operator

$$\frac{1}{c^2}\frac{\partial^2}{\partial t^2} - \nabla^2 , \tag{2.33}$$

that appears in Eqs. (2.9) and (2.10), is invariant with respect to (2.29), (2.30), (2.31), (2.32). Taking into account the rules of differential calculus we find

$$\frac{\partial}{\partial x} = \frac{\partial x'}{\partial x}\frac{\partial}{\partial x'} + \frac{\partial t'}{\partial x}\frac{\partial}{\partial t'} = \gamma\frac{\partial}{\partial x'} - \gamma\frac{\beta}{c}\frac{\partial}{\partial t'} , \tag{2.34}$$

$$\frac{\partial}{\partial y} = \frac{\partial}{\partial y'} , \tag{2.35}$$

$$\frac{\partial}{\partial z} = \frac{\partial}{\partial z'} , \tag{2.36}$$

$$\frac{\partial}{\partial t} = \frac{\partial x'}{\partial t}\frac{\partial}{\partial x'} + \frac{\partial t'}{\partial t}\frac{\partial}{\partial t'} = -\gamma v\frac{\partial}{\partial x'} + \gamma\frac{\partial}{\partial t'} . \tag{2.37}$$

It follows that

$$\frac{\partial^2}{\partial x^2} = \gamma^2\frac{\partial^2}{\partial (x')^2} + \gamma^2\frac{\beta^2}{c^2}\frac{\partial^2}{\partial (t')^2} - 2\gamma^2\frac{\beta}{c}\frac{\partial^2}{\partial x'\partial t'} , \tag{2.38}$$

$$\frac{\partial^2}{\partial y^2} = \frac{\partial^2}{\partial (y')^2} , \tag{2.39}$$

$$\frac{\partial^2}{\partial z^2} = \frac{\partial^2}{\partial (z')^2} , \tag{2.40}$$

$$\frac{\partial^2}{\partial t^2} = \gamma^2 v^2\frac{\partial^2}{\partial (x')^2} + \gamma^2\frac{\partial^2}{\partial (t')^2} - 2\gamma^2 v\frac{\partial^2}{\partial x'\partial t'} . \tag{2.41}$$

Combining Eqs. (2.38) and (2.41) we obtain

$$\frac{1}{c^2}\frac{\partial^2}{\partial t^2} - \frac{\partial^2}{\partial x^2} = \frac{\gamma^2}{c^2}(1-\beta^2)\frac{\partial^2}{\partial (x')^2} - \gamma^2\left(1-\frac{v^2}{c^2}\right)\frac{\partial^2}{\partial (x')^2} - 2\gamma^2\left(\frac{v}{c^2} - \frac{\beta}{c}\right)\frac{\partial^2}{\partial x'\partial t'} \tag{2.42}$$

which gives

$$\frac{1}{c^2}\frac{\partial^2}{\partial t^2} - \frac{\partial^2}{\partial x^2} = \frac{1}{c^2}\frac{\partial^2}{\partial (t')^2} - \frac{\partial^2}{\partial (x')^2} \tag{2.43}$$

only if $\beta = v/c$ and $\gamma^2 = 1 - \beta^2 = 1 - v^2/c^2$. These are the parameters of the so-called Lorentz transformations.

2.2   Lorentz Transformations

We have just proved that the d'Alembert equations of the electromagnetic waves (2.9) and (2.10) are not invatiant with respect to the Galilei transformations (2.25)–(2.28) but are instead invariant with respect to the Lorentz transformations

$$x' = \frac{x - vt}{\sqrt{1 - \frac{v^2}{c^2}}}, \tag{2.44}$$

$$y' = y, \tag{2.45}$$

$$z' = z, \tag{2.46}$$

$$t' = \frac{t - vx/c^2}{\sqrt{1 - \frac{v^2}{c^2}}}. \tag{2.47}$$

It is possible to prove that also Maxwell's equations (2.1)–(2.4) are invariant with respect to these Lorentz transformations. Moreover, it is straightforward to verify that, under the condition

$$\frac{v}{c} \ll 1, \tag{2.48}$$

the Lorentz transformations reduce to the Galilei transformations. Thus, if the relative velocity v of the two reference systems is much smaller than the speed of light c, Lorentz and Galilei transformations are practically equivalent.

2.2.1   Thought Experiment with Light Bulb

In 1905 Albert Einstein recognized the strict connection between the invariance of the speed of light and the Lorentz transformations. Here we analyze a Gedankenexperiment (thought experiment) that was discussed in 1916 by Einstein to explain in a simple way this connection.

Let us consider a small light bulb that at time zero is turned on emitting light in all directions. Assume that the emission of light is uniform. Consider a stationary reference system O with origin in the center of the bulb. According to this reference system a point P on the surface of the sphere of light emitted by the bulb will have coordinates $\mathbf{r} = (x, y, z)$ which satisfy the equation

$$|\mathbf{r}| = ct, \tag{2.49}$$

where c is the speed of light and t the time measured by the system reference system O. This formula can be rewritten as

$$\sqrt{x^2 + y^2 + z^2} = ct \tag{2.50}$$

That is, by squaring it we get

$$x^2 + y^2 + z^2 = c^2 t^2 . \tag{2.51}$$

This is the equation of a sphere of radius r that grows with time with the law $r = ct$.

Let us now consider another reference system O' such that the coordinates $\mathbf{r}' = (x', y', z')$ of the point P with respect to this reference system reference system are related to those of the reference system O by the generic transformations (2.29)–(2.32), where t' is the time measured by the reference system O'. Again, in these generic transformations, γ and β are dimensionless parameters that must be determined. According to the reference system O', a point P on the surface of the sphere of light emitted by the bulb will have coordinates $\mathbf{r}' = (x', y', z')$ that satisfy the equation

$$|\mathbf{r}'| = c' t' , \tag{2.52}$$

or

$$(x')^2 + (y')^2 + (z')^2 = (c')^2 (t')^2 . \tag{2.53}$$

Einstein's crucial assumption, based on the experiment of Michelson-Morley experiment and earlier work by Poincare and others, is

$$c = c' . \tag{2.54}$$

Therefore we have

$$0 = x^2 + y^2 + z^2 - c^2 t^2 = (x')^2 + (y')^2 + (z')^2 - c^2 (t')^2 . \tag{2.55}$$

Inserting in this expression the generic space-time coordinate transformation given by Eqs. (2.29)–(2.32), after some algebraic calculations we obtain

$$\gamma = \frac{1}{\sqrt{1 - \frac{v^2}{c^2}}} \qquad \text{and also} \qquad \beta = \frac{v}{c} . \tag{2.56}$$

Thus, in conclusion, we have found that correct transformations which ensure the invariance of the speed of light are nothing else than the Lorentz transformations.

2.3 Einstein Postulates

This research activity on invariant transformations was summarized by Albert Einstein, who proposed to adopt two suggestive postulates:

P1: The laws of physics are the same for all inertial reference systems.

P2: The speed of light in vacuum is the same in all inertial reference systems.

It is important to stress that a reference system is called *inertial* if the first principle of dynamics is valid for it: a material point not subject to external forces measured in this reference system is either at rest or moving at a constant speed. In other words, two reference systems are inertial if their relative velocity remains constant.

From his two postulates Einstein deduced that the laws of physics must be invariant with respect to the Lorentz transformations but the laws of Newtonian mechanics (which are not) must be modified.

2.4 Relativistic Kinematics

Einstein developed a new mechanics, the relativistic mechanics, which relates back to Newtonian mechanics when the velocities of massive particles are much smaller than the speed of light c. As known, mechanics can be divided into two parts: the kinematics and dynamics. We begin here to analyze the relativistic kinematics.

2.4.1 Length Contraction

One of the surprising results of relativistic kinematics is the contraction of length: the length L of a rod measured by an observer (measuring instrument) moving at the velocity v with respect to the rod given by

$$L = L_0\sqrt{1 - \frac{v^2}{c^2}}\,, \tag{2.57}$$

where L_0 is the proper length of the rod, i.e., the length measured by an observer for whom the rod is at rest (Fig. 2.1).

The contraction of lengths is easily verified using the first equation of Lorentz transformations:

$$x' = \frac{x - v t}{\sqrt{1 - \frac{v^2}{c^2}}}\,. \tag{2.58}$$

A finite variation of the space-time coordinates implies

$$\Delta x' = \frac{\Delta x - v\,\Delta t}{\sqrt{1 - \frac{v^2}{c^2}}}\,. \tag{2.59}$$

In our problem $\Delta x = L$ is the length of the bar with respect to the system of reference O. The length is measured by O instantaneously, i.e. with $\Delta t = 0$. Clearly $\Delta x' = L_0$

Fig. 2.1 Scaled length L/L_0 of the rod as a function of the scaled velocity v/c of the rod

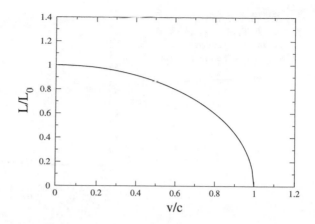

is instead the length of the bar measured at the reference system O' which is comoving with the bar. Therefore we have

$$L_0 = \frac{L}{\sqrt{1 - \frac{v^2}{c^2}}} \qquad (2.60)$$

or, equivalently, Eq. (2.57). Note that the length contraction appears only in the direction of the moving observer.

2.4.2 Time Dilation

Another surprising result of relativistic kinematics is the dilation of time: the time interval T of a clock measured by an observer moving at the speed v with respect to the clock is given by

$$T = \frac{T_0}{\sqrt{1 - \frac{v^2}{c^2}}}, \qquad (2.61)$$

where T_0 is the proper time of the clock, i.e. the time measured by an observer for whom the clock is at rest (Fig. 2.2).

The dilation of time is easily demonstrated by using the fourth equation of Lorentz transformations:

$$t' = \frac{t - \frac{v}{c^2} x}{\sqrt{1 - \frac{v^2}{c^2}}}. \qquad (2.62)$$

A finite variation of the space-time coordinates implies

Fig. 2.2 Scaled time interval T/T_0 of the clock as a function of the scaled velocity v/c of the clock

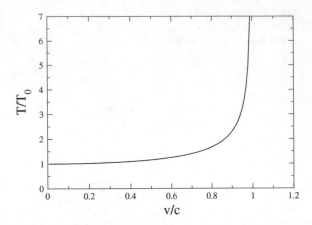

$$\Delta t' = \frac{\Delta t - \frac{v}{c^2}\Delta x}{\sqrt{1 - \frac{v^2}{c^2}}} \, . \tag{2.63}$$

In our problem $\Delta t = T_0$ is the time interval of a clock with respect to the reference system O, comoving with the clock, and measured with the clock at the same position, i.e. with $\Delta x = 0$. Clearly $\Delta t' = T$ is instead the time interval measured by the reference system O'. Therefore we obtain precisely Eq. (2.61).

2.4.3 Transformation of Velocities

Let us consider the Lorentz transformations given by Eqs. (2.44)–(2.47). An infinitesimal variation of the space-time coordinates implies

$$dx' = \frac{dx - v\, dt}{\sqrt{1 - \frac{v^2}{c^2}}} \, , \tag{2.64}$$

$$dy' = dy \, , \tag{2.65}$$

$$dz' = dz \, , \tag{2.66}$$

$$dt' = \frac{dt - v\, dx/c^2}{\sqrt{1 - \frac{v^2}{c^2}}} \, . \tag{2.67}$$

The velocity $\mathbf{V} = (V_x, V_y, V_z)$ of the particle P, as seen by the reference system O, is defined as

$$V_x = \frac{dx}{dt} \, , \qquad V_y = \frac{dy}{dt} \, , \qquad V_z = \frac{dz}{dt} \, , \tag{2.68}$$

while the velocity $\mathbf{V}' = (V_x', V_y', V_z')$ of the particle P, as seen by the reference system O', is instead defined as

$$V_x' = \frac{dx'}{dt'}, \qquad V_y' = \frac{dy'}{dt'}, \qquad V_z' = \frac{dz'}{dt'}. \qquad (2.69)$$

By using the differential Lorentz transformations (2.64)–(2.67) one derives the Lorentz transformations for the velocities

$$V_x' = \frac{V_x - v}{1 - \frac{v V_x}{c^2}}, \qquad (2.70)$$

$$V_y' = \frac{V_y \sqrt{1 - \frac{v^2}{c^2}}}{1 - \frac{v V_x}{c^2}}, \qquad (2.71)$$

$$V_z' = \frac{V_z \sqrt{1 - \frac{v^2}{c^2}}}{1 - \frac{v V_x}{c^2}}. \qquad (2.72)$$

These relativistic transformations of velocities are much more complicated than the relativistic transformations of positions. However, for $v \ll c$ and $V_x \ll c$ one gets

$$V_x' = V_x - v, \qquad (2.73)$$
$$V_y' = V_y, \qquad (2.74)$$
$$V_z' = V_z, \qquad (2.75)$$

that are the familiar Galilei transformations of velocities. On the other hand, setting $\mathbf{V} = (c, 0, 0)$ in Eqs. (2.70)–(2.72) it follows that $\mathbf{V}' = (c, 0, 0)$, which is fully consistent with the postulate P2 of Einstein about the invariance of the speed of light.

2.5 Relativistic Dynamics

The postulate P1 of Einstein says that laws of physics are invariant with respect to Lorentz transformations. However, the laws of Newtonian mechanics, which are invariant with respect to Galilei transformation, are not invariant with respect to Lorentz transformations. For this reason, Einstein concluded that the Newton law

$$\mathbf{F} = m\,\mathbf{a}, \qquad (2.76)$$

where $\mathbf{F}$ is the force acting on a particle of constant mass m moving with acceleration $\mathbf{a}$, must be modified. Einstein suggested the following relativistic Newton law

$$\mathbf{F} = \frac{d\mathbf{p}}{dt}, \qquad (2.77)$$

where

$$\mathbf{p} = \frac{m\,\mathbf{v}}{\sqrt{1 - \frac{v^2}{c^2}}}\,. \tag{2.78}$$

is the relativistic linear momentum of the particle of mass m moving with velocity $\mathbf{v}$. It is possible to prove that Eq. (2.77) equipped with Eq. (2.78) is indeed Lorentz invariant. It is important to stress that the relativistic linear momentum can also written as

$$\mathbf{p} = m_R\,\mathbf{v} \tag{2.79}$$

where

$$m_R = \frac{m}{\sqrt{1 - \frac{v^2}{c^2}}} \tag{2.80}$$

is the so-called relativistic mass of the particle. However, modern scientific books and papers do not use anymore the concept of relativistic mass.

It can be seen immediately that if $v \ll c$ the relativistic linear momentum (2.78) reduces to the familiar non-relativistic expression

$$\mathbf{p} = m\,\mathbf{v}\,, \tag{2.81}$$

and the relativistic Newton law reduces to the non-relativistic one, i.e. Eq. (2.76). On the other hand, if the particle is moving at the speed of of light, that is $v = c$, it follows that

$$\mathbf{p} = \frac{m\,\mathbf{v}}{0}\,. \tag{2.82}$$

This expression is mathematically ill-defined, unless one assumes that the mass m is zero. In this case

$$\mathbf{p} = \frac{\mathbf{0}}{0} \tag{2.83}$$

which is an indeterminate form of type $0/0$. This results somehow suggests that a particle moving at the speed of light must have zero mass. We will see that, indeed, the light is made of fotons, which are particles with zero mass.

2.5.1   Mechanical Work and Relativistic Energy

Also in relativistic mechanics the definition of mechanical work W_{AB} done by a particle to move from the point A to the point B is given by

$$W_{AB} = \int_A^B \mathbf{F} \cdot d\mathbf{r}\,. \tag{2.84}$$

We now show that the theorem which relates the mechanical work to the kinetic-energy difference is valid also in relativistic mechanics but the relativistic kinetic energy is different with respect to the non-relativistic one. By using the relativistic Eqs. (2.77) and (2.78) we obtain

$$
W_{AB} = \int_A^B \frac{d\mathbf{p}}{dt} \cdot d\mathbf{r} = \int_A^B d\mathbf{p} \cdot \frac{d\mathbf{r}}{dt} = \int_A^B d\mathbf{p} \cdot \mathbf{v}
$$

$$
= -\int_A^B \mathbf{p} \cdot d\mathbf{v} + [\mathbf{p} \cdot \mathbf{v}]_A^B = -\int_A^B \frac{m}{\sqrt{1 - \frac{v^2}{c^2}}} v\, dv + \left[\frac{mv^2}{\sqrt{1 - \frac{v^2}{c^2}}} \right]_A^B
$$

$$
= E_K(v_B) - E_K(v_A) = \Delta E_K , \tag{2.85}
$$

where

$$
E_K(v) = \frac{mc^2}{\sqrt{1 - \frac{v^2}{c^2}}} - mc^2 \tag{2.86}
$$

is the relativistic kinetic energy, such that

$$
E_K(v = 0) = 0 . \tag{2.87}
$$

The quantity

$$
E_R = mc^2 \tag{2.88}
$$

is called rest energy of the particle.

It is important to observe that the relativistic kinetic energy $E_K(v)$, given by Eq. (2.86), can be Taylor expanded around $v = 0$ giving

$$
E_K(v) = \frac{1}{2}mv^2 - \frac{mv^4}{8c^2} + \dots , \tag{2.89}
$$

which reduces to the familiar non-relativistic kinetic energy $mv^2/2$ if $v \ll c$.

The rest energy $E_R = mc^2$ is extremely important because it explicitly shows the *mass-energy equivalence*: all massive particles have an intrinsic energy directly related to their mass. One can then introduce the total relativistic energy of a free particle as

$$
E = E_R + E_K = \frac{mc^2}{\sqrt{1 - \frac{v^2}{c^2}}} . \tag{2.90}
$$

In general, the conservation of the total energy is a universal principle in Physics.

2.5.2 Relativistic Energy and Linear Momentum

Up to now we have discussed two relevant relativistic formulas for the linear momentum, Eq. (2.78), and for the energy, Eq. (2.90). Squaring the two expressions we immediately obtain

$$p^2 = \frac{m^2 v^2}{1 - \frac{v^2}{c^2}} \,, \tag{2.91}$$

$$E^2 = \frac{m^2 c^2}{1 - \frac{v^2}{c^2}} \,. \tag{2.92}$$

Then, after simple manipulations, we find that

$$E^2 - p^2 c^2 = m^2 c^4 \,, \tag{2.93}$$

namely

$$E = \sqrt{m^2 c^4 + p^2 c^2} \,. \tag{2.94}$$

This is a very nice results of Einstein's relativistic dynamics: the energy E of a particle in terms of its mass m and its linear momentum $p = |\mathbf{p}|$.

Clearly, if the particle has zero momentum, that is $p = 0$, then from Eq. (2.94) we get

$$E = mc^2 = E_R \,, \tag{2.95}$$

which is the previously introduced rest energy of the particle. If instead the particle has zero mass, that is $m = 0$, from Eq. (2.94) we obtain

$$E = c\,p \,, \tag{2.96}$$

which is the energy of a particle with zero mass. We have seen previously that the linear momentum $\mathbf{p}$ of a particle of zero mass (which must have velocity equal to the speed of light) is undetermined. It follows that its energy E is also undetermined. Despite this, the two indeterminate quantities are related to each other by the Eq. (2.96). It is now estabished that the elementary particles that constitute the light, the so-called photons, are particles with zero mass.

2.5.3 Non-relativistic Limit of the Energy

In general, for a particle with $p \neq 0$ and $m \neq 0$, Eq. (2.94) can be rewritten as

$$E = mc^2\sqrt{1 + \frac{p^2}{m^2c^2}}.$$ (2.97)

We can Taylor expanding the square root of Eq. (2.97) obtaining

$$E = mc^2 + \frac{p^2}{2m} - \frac{p^4}{8m^3c^2} + \dots.$$ (2.98)

This shows, another time, that the energy E is given by the sum of two contributions: the rest energy mc^2 and the relativistic kinetic energy

$$E_K = \frac{p^2}{2m} - \frac{p^4}{8m^3c^2} + \dots$$ (2.99)

which reduces to the non-relativistic kinetic energy $p^2/(2m)$ if $p \ll mc$.

As previously stressed, the formulas of relativistic dynamics have been confirmed by many experiments with interacting relativistic particles. Among them there are the particle-physics experiments routinely performed in the colliders of the European Organization for Nuclear Research (CERN) and the Fermi National Accelerator Laboratory (Fermilab). In the next chapters we will see that the relativistic theory plays a crucial role also in atomic, molecular, and optical physics.

2.6 Basic Concepts of General Relativity

The theory of general relativity, formulated by Albert Einstein in 1915, is the modern description of gravitational phenomena. This theory is a coherent account of gravity as a geometric characteristic of four-dimensional spacetime, that generalizes special relativity and refines Newton's law of universal gravitation.

2.6.1 Spacetime Interval

We have seen that in special relativity space and time are strictly related. It is then quite natural to introduce a four-dimensional spacetime point $x = (x^0, x^1, x^2, x^3)$ to represent the spacetime coordinates of a particle seen by a frame of reference. The four-position x is also called spacetime event. The time component x^0 is defined as $x^0 = ct$ with c the speed of light in vacuum and t the time coordinate. The space components x^1, x^2, x^3 are nothing else than the familiar coordinates of the three-dimensional position vector $\mathbf{r} = (x^1, x^2, x^3)$. A generic component of the spacetime point x is denoted by x^μ with $\mu = 0, 1, 2, 3$. This x^μ is called contravariant component of the four-vector x. Formally, one can write

$$x^\mu = \begin{pmatrix} x^0 \\ x^1 \\ x^2 \\ x^3 \end{pmatrix} \tag{2.100}$$

By using these notations, the Lorentz transformations (2.44)–(2.47) become

$$x^{0'} = \frac{x^0 - \beta x^1}{\sqrt{1 - \beta^2}}, \tag{2.101}$$

$$x^{1'} = \frac{x_1 - \beta x_0}{\sqrt{1 - \beta^2}}, \tag{2.102}$$

$$x^{2'} = x^2, \tag{2.103}$$

$$x^{3'} = x^3, \tag{2.104}$$

with $\beta = v/c$. One can then introduce the infinitesimal spacetime interval ds such that

$$ds^2 = (dx^0)^2 - (dx^1)^2 - (dx^2)^2 - (dx^3)^2 = c^2 dt^2 - d\mathbf{r}^2 = c^2 dt^2 \left(1 - \frac{v^2}{c^2}\right), \tag{2.105}$$

where v is here defined as $v = |\mathbf{v}| = |d\mathbf{r}/dt|$. Quite remarkably this spacetime interval is invariant for inertial frames. In fact, by using the Lorentz transformations (2.101)–(2.104) we have

$$\begin{aligned}
(ds')^2 &= (dx^{0'})^2 - (dx^{1'})^2 - (dx^{2'})^2 - (dx^{3'})^2 \\
&= \frac{(dx^0 - \beta dx^1)^2}{(1 - \beta^2)} - \frac{(dx^1 - \beta dx^0)^2}{(1 - \beta^2)} - (dx^2)^2 - (dx^3)^2 \\
&= \frac{(1 - \beta^2)\left((dx^0)^2 - (dx^1)^2\right)}{(1 - \beta^2)} - (dx^2)^2 - (dx^3)^2 \\
&= ds^2. \tag{2.106}
\end{aligned}$$

The infinitesimal spacetime interval (2.105) can be also written is a more formal way as

$$ds^2 = \eta_{\mu\nu} \, dx^\mu \, dx^\nu, \tag{2.107}$$

where $\eta_{\mu\nu}$ is the *Minkowski metric tensor*, dx^μ is an infinitesimal variation of x^μ and repeated indices means the summation over them. The metric tensor $\eta_{\mu\nu}$ is the element of a 4×4 matrix. Formally, we can write

$$\eta_{\mu\nu} = \begin{pmatrix} 1 & 0 & 0 & 0 \\ 0 & -1 & 0 & 0 \\ 0 & 0 & -1 & 0 \\ 0 & 0 & 0 & -1 \end{pmatrix}. \tag{2.108}$$

It is important to stress that the spacetime interval is not invariant for non-intertial frames, for instance if one frame of reference is accelerating. Thus, in full generality, for a non-intertial frame the infinitesimal spacetime interval ds reads

$$ds^2 = g_{\mu\nu}(x)\, dx^\mu\, dx^\nu \,, \tag{2.109}$$

where $g_{\mu\nu}(x)$ is the local *spacetime metric tensor*. The metric tensor $g_{\mu\nu}(x)$ is the element of a 4×4 matrix. Formally, we can write

$$g_{\mu\nu}(x) = \begin{pmatrix} g_{00}(x) & g_{01}(x) & g_{02}(x) & g_{03}(x) \\ g_{10}(x) & g_{11}(x) & g_{12}(x) & g_{13}(x) \\ g_{20}(x) & g_{21}(x) & g_{22}(x) & g_{23}(x) \\ g_{30}(x) & g_{31}(x) & g_{32}(x) & g_{33}(x) \end{pmatrix} . \tag{2.110}$$

However the problem is a bit simpler because $g_{\mu\nu}(x)$ is symmetric tensor, i.e. $g_{\mu\nu}(x) = g_{\nu\mu}(x)$. This implies that $g_{\mu\nu}(x)$ has 10 independent components.

Given the infinitesimal spacetime interval ds one can introduce the infinitesimal proper time $d\tau = c\, ds$, which is an invariant quantity in the case of intertial frames. A finite proper time interval $\Delta\tau$ is then given by

$$\Delta\tau = \int_{\mathcal{P}} d\tau = \int_{\mathcal{P}} \frac{ds}{c} = \int_{\mathcal{P}} \frac{1}{c} \sqrt{g_{\mu\nu}(x)\, dx^\mu\, dx^\nu} \tag{2.111}$$

where $\mathcal{P}$ is a worldline, i.e. a spacetime path. This $\Delta\tau$ is time interval measured by a clock that is at rest with the worldline.

2.6.2   Curved Manifolds

It is important to observe that Eq. (2.109) is nothing else than the extension of the notion of infinitesimal spatial interval introduced by Bernhard Riemann for curved manifolds. In other words, the spacetime is treated as a four-dimensional manifold characterized by the local metric tensor $g_{\mu\nu}(x)$. At the end of XIX century, the motion on curved manifolds was deeply anayzed by Elwin Bruno Christoffel, Gregorio Ricci-Curbastro, and Tullio Levi-Civita developing the so-called tensor calculus. Within the framework of the tensor calculus, the covariant component x_μ of the four-vector x is defined as $x_\mu = g_{\mu\nu}(x)\, x^\nu$, where again repeated indices means the summation over them. Moreover, $g^{\mu\nu}(x)$ represents a generic element of the 4×4 inverse matrix obtained from the matrix of $g_{\mu\nu}(x)$.

One of the main results of differential geometry and tensor calculus is the following one. In the absence of external forces, for a particle of local coordinates $x(\tau) = (x^0(\tau), x^1(\tau), x^2(\tau), x^3(\tau))$ and some time τ (which can be the proper time in the case of general relativity), that is moving on the curved manifold characterized by the local metric tensor $g_{\mu\nu}(x)$, the acceleration is given by

$$\frac{d^2 x^\alpha}{d\tau^2} = -\Gamma^\alpha_{\mu\nu} \frac{dx^\mu}{d\tau} \frac{dx^\nu}{d\tau} , \qquad (2.112)$$

where

$$\Gamma^\alpha_{\mu\nu}(x) = \frac{1}{2} g^{\alpha\beta}(x) \left(\frac{\partial}{\partial x^\nu} g_{\beta\mu}(x) + \frac{\partial}{\partial x^\mu} g_{\beta\nu}(x) - \frac{\partial}{\partial x^\beta} g_{\mu\nu}(x) \right) \qquad (2.113)$$

is the Christoffel symbol. Thus, the familiar acceleration $d^2 x^\alpha / d\tau^2$ of the particle is not zero because the particle is constrained to move on the curved manifold, and the Christoffel symbol $\Gamma^\alpha_{\mu\nu}$ takes into account the presence of the constraining manifold. In the case of a x-independent metric tensor the Christoffel symbols are zero and the acceleration of the particle is zero.

2.6.3 Equivalence Principle and Einstein Equations

Einstein noticed that a person in a freely falling elevator would experience no apparent weight; that is, the inertial mass m_I of a test particle is equal to its gravitational mass m_G. In fact, the Newton law of the non-relativistic dyamics of this test particle under the effect of the gravitational force due to another particle of gravitational mass M_G is given by

$$m_I \frac{d^2 \mathbf{r}}{dt^2} = -G \frac{m_G M_G}{|\mathbf{r}|^2} , \qquad (2.114)$$

where $G = 6.7 \times 10^{-11} \, \text{m}^3 \, \text{kg}^{-1} \, \text{s}^{-2}$ is the gravitational constant and $\mathbf{r}(t)$ is the position vector of the test particle measured with respect to the other particle. However, experimentally (Galileo Galilei, 1590; Lorand Eötvös 1922) one finds that $m_G = m_I$, and also $M_G = M_I$. Consequently, we can write

$$\frac{d^2 \mathbf{r}}{dt^2} = -G \frac{M}{|\mathbf{r}|^2} , \qquad (2.115)$$

where we set $M = M_G = M_I$ and $m = m_G = m_I$ for the two masses. For these reasons, Einstein formulated new postulates, in addition to the ones of the special relativity:

P3: The laws of physics in a gravitational field are identical to those of a local accelerating frame.

P4: The laws of physics must be expressed by identical equations relative to all other systems, whichever way they are moving.

The postulate P3 is called equivalence principle, while the postulate P4 is called principle of general covariance. In other words, Einstein supposed that the gravitational effects must be all encoded into the spacetime metric tensor $g_{\mu\nu}(x)$. According to the postulate P3, in the absence of any energy or momentum at the spacetime point x, for an inertial frame of reference the spacetime interval (2.109) is such that $g_{\mu\nu}(x) = \eta_{\mu\nu}$, as in Eq. (2.107). Instead, if there is some kind of energy or momentum at the spacetime point x, the metric tensor $g_{\mu\nu}(x)$ is much more complicated.

Einstein, after several trials and errors, and taking into account the postulates P3 and P4, found that $g_{\mu\nu}(x)$ must satisfy the following equations

$$R_{\mu\nu}(x) - \frac{1}{2}R(x)\, g_{\mu\nu}(x) = \frac{8\pi G}{c^4}\, T_{\mu\nu}(x)\ . \tag{2.116}$$

We do not derive these equations, which were obtained by Einstein in 1915 with the crucial technical help of David Hilbert. In these equations G is the gravitational constant and c is the speed of light in vacuum. The quantity

$$R_{\mu\nu}(x) = \frac{\partial}{\partial x^\alpha}\Gamma^\alpha_{\nu\mu}(x) - \frac{\partial}{\partial x^\mu}\Gamma^\alpha_{\alpha\mu}(x) + \Gamma^\alpha_{\alpha\beta}(x)\,\Gamma^\beta_{\nu\mu}(x) - \Gamma^\alpha_{\nu\beta}\,\Gamma^\beta_{\alpha\mu}(x) \tag{2.117}$$

is the Ricci tensor, which is a symmetric tensor. Instead

$$R(x) = q^{\mu\nu}(x)\, R_{\mu\nu}(x) \tag{2.118}$$

is the Ricci scalar, and $T_{\mu\nu}(x)$ is the stress-energy tensor, which is also a symmetric tensor, i.e. $T_{\mu\nu}(x) = T_{\nu\mu}(x)$. Both the Ricci tensor $R_{\mu\nu}(x)$ and the Ricci scalar $R(x)$ depend on the metric tensor $g_{\mu\nu}(x)$ in a highly nonlinear and differential way through the Christoffel symbol $\Gamma^\alpha_{\mu\nu}(x)$, Eq. (2.113), which is indeed a crucial quantity of differential geometry. Ricci is the abbreviation for Ricci-Curbastro. The stress-energy tensor $T_{\mu\nu}(x)$ is the element of a 4×4 matrix, and it contains all the informations about the energy-momentum which resides at x due to presence of some source. As an example, for a very simple relativistic fluid with local mass density $\rho(x)$, and local relativistic four-velocity $u(x) = (u^0(x), u^1(x), u^2(x), u^3(x))$ with generic component $u^\mu(x) = dx^\mu/ds$, the stress-energy tensor reads

$$T_{\mu\nu}(x) = \rho(x)c^2 u_\mu(x)\, u_\nu(x)\ , \tag{2.119}$$

where $u_\mu(x) = g_{\mu\nu}(x)\, u^\nu(x)$ are the so-called covariant components of the relativistic four-velocity $u(x)$, which is an adimensional quadrivectorial quantity. In the case of a Minkoswki spacetime, according to Eq. (2.105) we have $ds = cdt\sqrt{1 - v^2/c^2}$ and the time-time component $T_{00}(x)$ of the stress-energy tensor becomes $T_{00}(x) = \rho(x)c^2/(1 - v^2/c^2)$, which is the energy density of this very simple relativistic fluid.

2.6.4   Non-Relativistic Limit of General Relativity

The partial differential equations (2.116) connect the local spacetime metric tensor $g_{\mu\nu}(x)$ with the local stress-tensor $T_{\mu\nu}(x)$. In general, the problem of solving Eq. (2.116), is a quite difficult task. However, the Einstein equations have been formulated in such a way that they reduce to the familiar gravitational theory of Newton in the non-relativistic regime, i.e. when the three-dimensional velocity $\mathbf{v} = d\mathbf{r}/dt$ is small compared to the speed of light c. In this limit the stress-energy tensor of our very simple fluid becomes

$$T_{\mu\nu}(x) = \begin{pmatrix} \rho(\mathbf{r})c^2 & 0 & 0 & 0 \\ 0 & 0 & 0 & 0 \\ 0 & 0 & 0 & 0 \\ 0 & 0 & 0 & 0 \end{pmatrix}, \tag{2.120}$$

where $\rho(\mathbf{r}) = \rho(x_0 = 0, x^1, x^2, x^3)$. Moreover, in the same Newtonian limit, from Eq. (2.116) after some calculations one finds

$$g_{\mu\nu}(x) = \begin{pmatrix} 1 + 2\phi(\mathbf{r})/c^2 & 0 & 0 & 0 \\ 0 & -1 & 0 & 0 \\ 0 & 0 & -1 & 0 \\ 0 & 0 & 0 & -1 \end{pmatrix} \tag{2.121}$$

for the local spacetime metric tensor, where the scalar field $\phi(\mathbf{r})$ satisfies the equation

$$\nabla^2 \phi(\mathbf{r}) = -4\pi G\, \rho(\mathbf{r}), \tag{2.122}$$

that is the Poisson equation for the gravitational potential $\phi(\mathbf{r})$ generated by a mass density distribution $\rho(\mathbf{r})$. This equation can be solved and gives

$$\phi(\mathbf{r}) = -4\pi G \int \frac{\rho(\mathbf{r}')}{|\mathbf{r} - \mathbf{r}'|}\, d^3\mathbf{r}'. \tag{2.123}$$

Considering a uniform density $\rho = M/V$ with total mass M inside a sphere of volume V, outside the sphere we get

$$\phi(\mathbf{r}) = -G\frac{M}{|\mathbf{r}|}, \tag{2.124}$$

which is the familiar Newtonian expression of the gravitational field generated by a mass M. In the non-relativistic limit, the formula (2.112) for the acceleration of a particle constrained to move on the curved spacetime with metric tensor given by Eq. (2.121) reads

$$\frac{d^2\mathbf{r}}{dt^2} = -\nabla\phi(\mathbf{r}), \tag{2.125}$$

which is exactly the Newton law for the non-relativistic dynamics of a particle under the action of the gravitational potential $\phi(\mathbf{r})$, i.e. Eq. (2.115).

2.6.5 Predictions of General Relativity

To conclude this section, we emphasize that many important predictions based on the Einstein equations of general relativity have been confirmed experimentally. The most famous is the explanation of the precession of the perihelion of Mercury. There are, however, very recent relevant achievements. Among them:
(i) the accelerating expansion of the Universe (Nobel Prize in Physis 2011);
(ii) the existence gravitational waves (Nobel Prize in Physics 2017);
(iii) the existence of black holes (Nobel Prize in Physics 2020).

Quantum mechanics and quantum field theory are used to explain three forces (electromagnetic, nuclear weak, and nuclear strong) of the four fundamental forces of Nature. Albert Einstein's general theory of relativity, which is formulated within the wholly distinct framework of classical physics, forms the basis for our current understanding of gravity, the fourth force. Nowadays, a very active field of research is quantum gravity. Quantum gravity aims to describe gravity using quantum mechanics principles in situations where quantum effects are important, such as in the vicinity of black holes or other compact astrophysical objects with strong gravitational effects (e.g. neutron stars).

Further Reading

Excellent books about special relativity are:
Landau, L.D., Lifshitz, E.M.: The Classical Theory of Fields. Pergamon (1980).
Rindler, W.: Introduction to Special Relativity. Oxford University Press (1991).
Relevant historical papers on special relativity are:
Michelson, A.A., Morley, E.W.: Am. J. Sci. **34**, 333 (1887).
Michelson, A.A., Morley, E.W.: Am. J. Sci. **34**, 427 (1887).
Poincare, H.: Revue de Metaphysique et de Morale **6**, 1 (1898).
Fitzgerald, G.F.: Science **13**, 390 (1889).
Lorentz, H.A.: Proc. R. Netherlands Acad. Arts Sci. **6**, 809 (1904).
Einstein, A.: Ann. der Physik **17**, 891 (1905).
An introductory book of general relativity is:
Ryder, L.: Introduction to General Relativity. Cambridge University Press (2009).
Relevant historical papers on general relativity are:
Hilbert, D.: Nachrichten von der Gesellschaft der Wissenschaften zu Gottingen, Mathematisch-Physikalische Klasse **3**, 395 (1915).
Einstein, A.: Sitzungsberichte der Preussischen Akademie der Wissenschaften zu Berlin, Part 2, 844 (1915).

Chapter 3
Quantum Properties of Light

In this chapter we discuss the three main empirical phenomena which, for the first time, emphasized the quantum nature of light. These phenomena are the black-body radiation, the photoelectric effect, and the Compton effect. The main idea is that light is made of elementary massless particles called photons. This idea, due to Planck, Einstein, and Compton, is now well established experimentally and it is the heart of the flourishing field of quantum optics.

3.1 Black-Body Radiation

Historically the beginning of quantum mechanics is placed in 1900 when Max Planck found that the experimental results of the electromagnetic spectrum emitted by a black body are explained under the assumption that the energy of the radiation is quantized.

A black body is a hypothetical physical body that completely absorbs all the oncoming electromagnetic radiation, regardless of frequency or incidence angle. The word black is due to the fact that this body absorbs all colors of light. However, as any hot body, the black body emits its own electromagnetic radiation to reach the thermal equilibrium with the external environment. A small hole in a big insulated container (cavity), with internal walls that are opaque to radiation, is a good approximation of the ideal black body. Any light that enters the hole is reflected or absorbed by the body's internal surfaces and is unlikely to re-emerge, making it a near-perfect absorber. Instead, the electromagnetic radiation outcoming from the hole, that is at thermal equilibrium with the walls of the cavity, is the black body radiation.

In 1898 Otto Lummer and Ferdinand Kurlbaum analyzed experimentally the electromagnetic radiation coming from the small hole of the cavity. By definition, $\rho(\nu)$ is such that

© The Author(s), under exclusive license to Springer Nature Switzerland AG 2022
L. Salasnich, *Modern Physics*, UNITEXT for Physics,
https://doi.org/10.1007/978-3-030-93743-0_3

Fig. 3.1 Black-body
radiation: Planck's law of the
energy density per unit
frequency, $\rho(\nu)$, for three
values of the temperature T

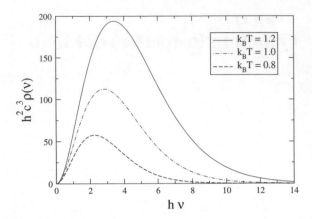

$$\frac{E}{V} = \int_0^{+\infty} \rho(\nu)\, d\nu \,, \tag{3.1}$$

where E is the total energy of the electromagnetic radiation inside a volume V.
Experimentally one finds that the function $\rho(\nu)$ is such that

$$\int_0^{+\infty} \rho(\nu)\, d\nu = a\, T^4 \,, \tag{3.2}$$

which is Stefan-Boltzmann's law with

$$a = 7.56 \cdot 10^{-16} \quad \text{J}/(\text{m}^3 \times \text{K}^4) \tag{3.3}$$

the so-called radiation constant. Moreover, one finds that the frequency ν_{max} of the
maximum of the curve grows by increasing the temperature (Wien's law).

As we will see in detail, Planck derived the expression (Fig. 3.1)

$$\rho(\nu) = \frac{8\pi}{c^3} \nu^2 \frac{h\nu}{e^{\frac{h\nu}{k_B T}} - 1} \tag{3.4}$$

for the density of electromagnetic energy per unit frequency $\rho(\nu)$ emitted by the
body at temperature T, with k_B the Boltzmann constant and c the speed of light in
vacuum. This formula, known as Planck's law of black-body radiation, is in very
good agreement with the experimental data. The constant h, derived by interpolation
of experimental data, results to be

$$h = 6.63 \cdot 10^{-34}\, \text{J} \times \text{s} \,. \tag{3.5}$$

This parameter is called Planck constant. Note that often we use also the reduced Planck constant

$$\hbar = \frac{h}{2\pi} = 1.06 \cdot 10^{-34} \, \text{J} \times \text{s} \tag{3.6}$$

that is routinely called "hbar". As expected, taking into account Eq. (3.4) we obtain Stefan-Boltzmann's law

$$\int_0^{+\infty} \rho(\nu) \, d\nu = \frac{8\pi^5 k_B^4}{15c^3 h^3} T^4 \, , \tag{3.7}$$

and this means that the radiation constant a of Eq. (3.3) must be

$$a = \frac{8\pi^5 k_B^4}{15c^3 h^3} = \frac{\pi^2 k_B^4}{15c^3 \hbar^3} \, . \tag{3.8}$$

Also the Wien's law is perfectly reproduced by Planck's black-body radiation formula.

3.1.1 Derivation of Planck's Law

The crucial assumption made by Max Planck to derive the correct function $\rho(\nu)$ of the black-body radiation is that the energy $E(\nu, n)$ emitted by the walls of the black-body cavity is quantized as follows

$$E(\nu, n) = h\nu \, n \, , \tag{3.9}$$

where h is Planck's constant, ν is the frequency of the radiation, and $n \in \mathbb{N}$ is a natural number called *quantum number*. According to Planck, the average electromagnetic energy $\bar{E}(\nu)$ emitted with frequency ν by the cavity at temperature T is given by

$$\bar{E}(\nu) = \langle E(\nu, n) \rangle = \frac{\sum_{n=0}^{+\infty} E(\nu, n) \, e^{-\beta E(\nu,n)}}{\sum_{n=0}^{+\infty} e^{-\beta E(\nu,n)}} \, , \tag{3.10}$$

where $e^{-\beta E(\nu,n)}$ is Boltzmann's statistical factor with $\beta = 1/(k_B T)$. Quite remarkably, Eq. (3.10) can be explicitly calculated. In fact, $\bar{E}(\nu)$ can be rewritten as

$$\bar{E}(\nu) = -\frac{\partial}{\partial \beta} \ln \mathcal{Z}_1 \tag{3.11}$$

where

$$\mathcal{Z}_1 = \sum_{n=0}^{+\infty} e^{-\beta h\nu \, n} \tag{3.12}$$

is the single-mode (single-frequency) partition function. By using the geometric series results, i.e. $\sum_{n=0}^{+\infty} x^n = 1/(1-x)$ for $|x| < 1$, we immediately obtain

$$\mathcal{Z}_1 = \frac{1}{e^{-\beta h\nu} - 1} . \tag{3.13}$$

Inserting this expression in Eq. (3.11) we get

$$\bar{E}(\nu) = \frac{h\nu}{e^{\frac{h\nu}{k_B T}} - 1} = h\nu\,\bar{n}(\nu) , \tag{3.14}$$

where

$$\bar{n}(\nu) = \frac{1}{e^{\frac{h\nu}{k_B T}} - 1} \tag{3.15}$$

is the thermal average number of quanta with frequency ν.

Let us now consider the dispersion relation

$$\nu = \frac{c}{\lambda} = \frac{ck}{2\pi} \tag{3.16}$$

and the fact that infinite wavevectors $\mathbf{k}$ correspond to the same wavenumber $k = |\mathbf{k}|$ and, consequently, to the same frequency ν. The total energy of the electromagnetic radiation is then given by

$$E = 2 \sum_{\mathbf{k}} \bar{E}\left(\nu = \frac{ck}{2\pi}\right) , \tag{3.17}$$

where the factor 2 takes into account the two possible polarization of the light. In the continuum limit, where $\sum_{\mathbf{k}} \rightarrow V \int d^3\mathbf{k}/(2\pi)^3$, we obtain

$$\frac{E}{V} = 2 \int \frac{d^3\mathbf{k}}{(2\pi)^3} \bar{E}\left(\nu = \frac{ck}{2\pi}\right) = \int_0^{+\infty} D_1(\nu)\,\bar{E}(\nu)\,d\nu = \int_0^{+\infty} \rho(\nu)\,d\nu , \tag{3.18}$$

where

$$D_1(\nu) = 2 \int \frac{d^3\mathbf{k}}{(2\pi)^3} \delta\left(\nu - \frac{ck}{2\pi}\right) = \frac{1}{\pi^2} \int_0^{+\infty} dk\, k^2 \delta\left(\nu - \frac{ck}{2\pi}\right) = \frac{8\pi}{c^3}\nu^2 \tag{3.19}$$

is the single-mode density of states per volume. Thus, the energy density $\rho(\nu)$ per unit of frequency can be expressed as

$$\rho(\nu) = D_1(\nu)\,\bar{E}(\nu) , \tag{3.20}$$

where $D_1(\nu)$ is the single-particle density of the states that have frequency ν whereas $\bar{E}(\nu)$ is given by Eq. (3.14). Explicity, we have

$$\rho(\nu) = \frac{8\pi}{c^3}\nu^2 \frac{h\nu}{e^{\frac{h\nu}{k_B T}} - 1}, \tag{3.21}$$

that is exactly the formula of Planck. In the regime $h\nu \ll k_B T$ of high temperatures we have

$$e^{\frac{h\nu}{k_B T}} \approx 1 + \frac{h\nu}{k_B T} \tag{3.22}$$

and consequently

$$\rho(\nu) \approx \frac{8\pi}{c^3}\nu^2 k_B T, \tag{3.23}$$

which is the Rayleigh-Jeans law: very good at low frequencies but unreliable at high frequencies (*ultraviolet catastrophe*). The Rayleigh-Jeans law was derived the same years of the Planck law by John William Strutt (Rayleigh) and James Jeans on the basis of classical statistical mechanics, i.e. using Eq. (3.20) with $D_1(\nu)$ given by Eq. (3.19) and $\bar{E}(\nu)$ given by the classical results $\bar{E}(\nu) = k_B T$ instead of the quantum one, Eq. (3.14). We underline that the classical result can be obtained assuming that the number n of Eq. (3.10) is real, i.e.

$$\bar{E}(\nu) = \langle E(\nu, n) \rangle = \frac{\int_0^{+\infty} E(\nu, n) \, e^{-\beta E(\nu,n)} \, dn}{\int_0^{+\infty} e^{-\beta E(\nu,n)} \, dn} = \frac{\int_0^{+\infty} h\nu \, n \, e^{-\beta h\nu n} \, dn}{\int_0^{+\infty} e^{-\beta h\nu n} \, dn} = \frac{1}{\beta} = k_B T. \tag{3.24}$$

Thus, it is really crucial to impose that n is a natural number to get the Planck law.

At this point it is useful to perform a simple exercise. We want to determine the corresponding energy density per unit of wavelength, $\rho(\lambda)$, of the black-body radiation. The linear frequency ν is related to the wavelength λ by the expression

$$\lambda\nu = c. \tag{3.25}$$

In practice, $\nu = c\lambda^{-1}$, from which

$$d\nu = -c\lambda^{-2} d\lambda. \tag{3.26}$$

By changing variable in the integral of Eq. (3.1) the energy density becomes

$$\frac{E}{V} = \int_0^\infty \frac{8\pi^2}{\lambda^5} \frac{hc}{e^{\beta hc/\lambda} - 1} \, d\lambda, \tag{3.27}$$

and consequently the energy density per unit of wavelength reads

$$\rho(\lambda) = \frac{8\pi^2}{\lambda^5} \frac{hc}{e^{\beta hc/\lambda} - 1}, \tag{3.28}$$

such that

$$\frac{E}{V} = \int_0^\infty \rho(\lambda) \, d\lambda \, . \tag{3.29}$$

3.2   Photoelectric Effect

A few years after the formulation of Planck's law for the black body, in 1905, Albert
Einstein suggested that the electromagnetic radiation is composed of light quanta,
called photons, where the energy E of a single photon is given by

$$E = h\nu = \hbar\omega \, , \tag{3.30}$$

with h the Planck constant, $\hbar = h/(2\pi)$ the reduced Planck constant, ν the frequency,
and $\omega = 2\pi\nu$ the angular frequency. Einstein used the concept of photon to explain
the photoelectric effect, that is the emission of electrons from a metallic surface when
this surface is hit by an electromagnetic radiation.

The photoelectric effect was observed for the first time at the end of the nineteenth
century by Heinrich Hertz and Philipp von Lenard. In the photoelectric experiment
two metallic plates are placed at a finite distance from each other inside a vacuum
chamber. The two plates are connected by a conducting wire, where there is an
ammeter, that measures the electric current, and a battery (voltage generator) which
can be switched off. A monochromatic light of frequency ν is sent to one of the two
metal plates. If the voltage generator is turned off, the passage of an electric current
is observed only if the frequency of the monochromatic light exceeds a critical value,
that is only if

$$\nu > \nu_0 \, , \tag{3.31}$$

where experimentally ν_0 turns out to depend on the material properties of the metal
plate hit by the light, (Table 3.1). Surprisingly, if $\nu < \nu_0$ one does not observe electric
current even if the intensity of the incident light is very large.

If the voltage generator is switched on with inverted polarity, the critical frequency
required for measuring current flow is larger. In fact, the negative electrons extracted

Table 3.1 Critical frequency ν_c of the photoelectric effect for some metals

Material	Critical frequency ν_c (Hz)	Light color
Potassium (K)	5.43×10^{14}	Green
Sodium (Na)	5.51×10^{14}	Green
Calcium (Ca)	7.74×10^{14}	Violet
Copper (Cu)	1.08×10^{15}	Ultraviolet
Ferro (Fe)	1.12×10^{15}	Ultraviolet
Silver (Ag)	1.14×10^{15}	Ultraviolet

from the material are repelled by the excess of negative charges found on the other plate. In this configuration with reversed polarity, there exists a critical potential difference, called stopping electric potential V_0, above which there is no electric current.

3.2.1 Theoretical Explanation

Einstein used the concept of a photon to explain the photoelectric effect:

(i) light consists of photons of energy $h\nu$;
(ii) the radiation-matter interaction is, with good approximation, an interaction between single photon and single electron;
(iii) the single electrons in metals are bound to the metal by a binding energy W;
(iv) the photoelectric effect occurs only if the energy $h\nu$ of the single photon interacting with the single electron is greater than the binding energy W.

Einstein suggested that the kinetic energy E_K of an electron emitted from the surface of a metal after being irradiated is given by

$$E_K = h\nu - W \,, \tag{3.32}$$

where W is the metal work function (i.e., the minimum energy to extract the electron from the surface of the metal). In the case of a turned off voltage generator, the Einstein formula clearly implies that the critical frequency ν_0 of the incident radiation required to extract electrons from a metal is

$$\nu_0 = \frac{W}{h} \,. \tag{3.33}$$

Then ν_0 is the ratio of the extraction energy W and the Planck constant h.

Furthermore, according to Einstein, the stopping electric potential V_0 necessary to prevent the single electron from reaching the other plate must be equal to the kinetic energy E_K of the electron, that is

$$eV_0 = E_K \,, \tag{3.34}$$

where $q = -e$ the electric charge of the electron with $e = 1.6 \cdot 10^{19}$ C. From the Eqs. (3.32) and (3.34) we immediately find

$$V_0 = \frac{h}{e}\nu - \frac{W}{e} \,. \tag{3.35}$$

Thus, V_0 grows linearly with frequency ν and the slope of this straight line is the ratio h/e of two universal constants. These results are in very good agreement with the experimental data. Einstein received the Nobel Prize in Physics in 1921 with this motivation: "for his services to theoretical physics, and especially for his discovery of the law of the photoelectric effect".

3.3   Energy and Linear Momentum of a Photon

If the light is composed of elementary particles called photons, obviously these photons move at the speed c of light. Based on Einstein relativistic dynamics, a particle moving at the speed of light, that is with $v = c$, must have zero mass, that is $m = 0$. Moreover, in this case the energy E of the particle is related to its momentum $\mathbf{p}$ by the relation

$$E = c\, p,\tag{3.36}$$

where $p = |\mathbf{p}|$. On the other hand, we have seen that for the photon

$$E = h\,\nu,\tag{3.37}$$

where h is Planck's constant and ν is the frequency of the light. It follows that

$$p = \frac{E}{c} = \frac{h\,\nu}{c}.\tag{3.38}$$

Recalling that

$$\lambda\,\nu = c,\tag{3.39}$$

where λ is the wavelength of light, we get

$$p = \frac{h}{\lambda}.\tag{3.40}$$

This formula relates a corpuscular property of light, the linear momentum p, with an undulatory property of light, the wavelength λ.

To conclude this section, let us perform two simple exercises. In the first exercise we want to calculate the number of photon emitted in 4 s by a lamp of 10 W which radiates 1% of its energy as monochromatic light with wavelength $6000 \cdot 10^{-10}$ m (orange light). The energy of one photon of wavelength λ and linear frequency ν is given by

$$\epsilon = h\nu = h\frac{c}{\lambda},\tag{3.41}$$

where is Planck's constant. In our problem one gets

$$\epsilon = h\frac{c}{\lambda} = 6.62 \cdot 10^{-34} \text{ J} \times \text{s} \times \frac{3 \cdot 10^8 \text{ m/s}}{6 \cdot 10^3 \cdot 10^{-10} \text{ m}} = 3.3 \cdot 10^{-19} \text{ J}.\tag{3.42}$$

During the period $\Delta t = 4\,\text{s}$ the energy of the lamp with power $P = 10\,\text{W}$ is

$$E = P\,\Delta t = 10\,\text{J/s} \cdot 4\,\text{s} = 40\,\text{J} \,. \tag{3.43}$$

The radiation energy is instead

$$E_{rad} = E \cdot 1\% = E \cdot \frac{1}{100} = \frac{40}{100}\,\text{J} = 0.4\,\text{J} \,. \tag{3.44}$$

The number of emitted photons is then

$$N = \frac{E_{rad}}{\epsilon} = \frac{0.4\,\text{J}}{3.3 \cdot 10^{-19}\,\text{J}} = 1.2 \cdot 10^{18} \,. \tag{3.45}$$

The second exercise is a bit more complicated. On a photoelectric cell it arrives a beam of light with wavelength $6500 \cdot 10^{-10}\,\text{m}$ and energy 10^6 erg per second [1 erg = 10^{-7} J]. This energy is entirely used to produce photoelectrons. We want to calculate the intensity of the electric current which flows in the electric circuit connected to the photoelectric cell. The wavelength can be written as

$$\lambda = 6.5 \cdot 10^3 \cdot 10^{-10}\,\text{m} = 6.5 \cdot 10^{-7}\,\text{m} \,. \tag{3.46}$$

The energy of the beam of light can be written as

$$E = 10^6\,\text{erg} = 10^6 \cdot 10^{-7}\,\text{J} = 10^{-1}\,\text{J} \,. \tag{3.47}$$

The time interval is

$$\Delta t = 1\,\text{s} \,. \tag{3.48}$$

The energy of a single photon reads

$$\epsilon = h\frac{c}{\lambda} = 6.6 \cdot 10^{-34}\frac{3 \cdot 10^8}{6.5 \cdot 10^{-7}}\,\text{J} = 3 \cdot 10^{-19}\,\text{J} \,. \tag{3.49}$$

The number of photons is thus given by

$$N = \frac{E}{\epsilon} = \frac{10^{-1}\,\text{J}}{3 \cdot 10^{-19}\,\text{J}} = 3.3 \cdot 10^{17} \,. \tag{3.50}$$

If one photon produces one electron with electric charge $e = -1.6 \cdot 10^{-19}$ C, the intensity of electric current is easily obtained:

$$I = \frac{|e|N}{\Delta t} = \frac{1.6 \cdot 10^{-19}\,\text{C} \cdot 3.3 \cdot 10^{17}}{1\,\text{s}} = 5.5. \cdot 10^{-2}\,\text{amp} \,. \tag{3.51}$$

To conclude we observe that with wavelength $6500 \cdot 10^{-10}$ m the photoelectric effect is possible only if the work function of the sample is reduced, for instance by using an external electric field.

3.4   Compton Effect

In 1923 Arthur Compton studied the scattering of a beam of X-rays that passes through a small target of graphite. Experimental data from this experiment show that after collision (scattering) at a certain scattering angle θ the light becomes bichromatic with two wavelengths:

$$\lambda_1 \approx \lambda, \tag{3.52}$$
$$\lambda_2 = \lambda + \lambda_c \left(1 - cos(\theta)\right), \tag{3.53}$$

where λ is the wavelength of light before the scattering. The empirical constant $\lambda_c = 2.42 \cdot 10^{-12}$ m is now known as the Compton wavelength. Clearly, in the case $\theta = 0$ we have

$$\lambda_2 = \lambda, \tag{3.54}$$

for $\theta = \pi/2 = 90°$ we instead obtain

$$\lambda_2 = \lambda + \lambda_c, \tag{3.55}$$

while the maximum separation occurs in the case $\theta = \pi = 180°$ where

$$\lambda_2 = \lambda + 2\lambda_c. \tag{3.56}$$

3.4.1   Theoretical Explanation

Compton himself gave a reasonable theoretical explanation of his experiment. Compton's hypotheses are:

  (i)   light is composed of photons, particles with zero mass;
 (ii)   graphite is composed of atoms and free electrons, in both cases they are particles with mass m different from zero;
(iii)   in the scattering process every single photon interacts with a single particle of graphite;
(iv)   this diffusion process is an elastic collision and therefore the total energy and the total momentum are conserved.

Before the scattering, the energy E_X and the linear momentum $\mathbf{p}_X$ of the photon are

$$E_X = h \frac{c}{\lambda} , \tag{3.57}$$

$$\mathbf{p}_X = \left(\frac{h}{\lambda}, 0, 0 \right) . \tag{3.58}$$

The energy E_g and the linear momentum $\mathbf{p}_g$ of a particle of mass m of graphite are instead

$$E_g = mc^2 , \tag{3.59}$$

$$\mathbf{p}_g = (0, 0, 0) , \tag{3.60}$$

in a reference system where initially the particle is stationary.

After the collision the energy E'_X and the linear momentum $\mathbf{p}'_X$ of the photon are

$$E'_X = h \frac{c}{\lambda'} , \tag{3.61}$$

$$\mathbf{p}'_X = \left(\frac{h}{\lambda'} \cos(\theta), \frac{h}{\lambda'} \sin(\theta), 0 \right) . \tag{3.62}$$

The energy E'_g and the linear momentum $\mathbf{p}'_g$ of the particle of mass m of graphite are instead

$$E'_g = \sqrt{m^2 c^4 + (p'_g c)^2} , \tag{3.63}$$

$$\mathbf{p}'_g = (p'_g \cos(\phi), p'_g \sin(\phi), 0) , \tag{3.64}$$

where θ is the scattering angle of the photon while ϕ is the scattering angle of the particle.

Imposing the conservation of the total energy

$$E_X + E_g = E'_X + E'_g \tag{3.65}$$

and also of the total linear momentum

$$\mathbf{p}_X + \mathbf{p}_g = \mathbf{p}'_X + \mathbf{p}'_g , \tag{3.66}$$

after some algebric calculations one finds

$$\lambda' = \lambda + \frac{h}{mc} (1 - \cos(\theta)) . \tag{3.67}$$

In the case of photon-electron collision $m = m_e = 9.11 \cdot 10^{-31}$ kg and $h/(m_e c)$ turns out to be the wavelength Compton (of the electron). And so $\lambda' = \lambda_2 = \lambda + \lambda_c (1 - \cos(\theta))$. Instead, in the case of photon-atom collision $m = m_C = 19.94 \cdot 10^{-27}$ kg and $h/(m_C c)$ turns out to be extremely small such that $\lambda' = \lambda_1 \approx \lambda$.

3.5 Pair Production

A photon generating an electron-positron pair near a nucleus is often referred to as pair creation. The positron is a particle with the same mass m_e of the electron but with opposite charge $+e$. The process is often written as

$$\gamma \rightarrow e^- + e^+ , \tag{3.68}$$

where γ is the symbol of a photon (with frequency ν in the gamma-ray region), e^- is the symbol of the electron, and e^+ is the symbol of the positron. Because the energy must be conserved, the photon energy $h\nu$ must be greater than the total rest mass energy $2m_e c^2$ of the two particles generated for pair creation to occur, i.e.

$$h\nu > 2m_e c^2 = 2 \times 0.511 \text{ MeV} = 1.022 \text{ MeV} , \tag{3.69}$$

with $\nu_{min} = 2m_e c^2/h = 2.5 \times 10^{20}$ Hz the minimum frequency of the gamma-ray photon. Literally, the phonon vanishes and the electron-positron pair appears in its place. The process is constrained by the conservation of energy and linear momentum.

To satisfy the conservation of linear momentum, the photon must be near a nucleus, as an electron–positron pair created in free space cannot meet both energy and momentum conservation. All other conserved quantum numbers of the produced particles must add up to zero, resulting in opposite values for the created particles. For example, if one particle has a positive electric charge $+e$, the other must have the electric charge $-e$. The phenomenon of pair priduction was first observed by Patrick Blackett and Giovanni Occhialini in 1933 with a counter-controlled cloud chamber. Experiments show that the probability of pair creation increases with photon energy and also with the square of the atomic number (number of protons) of the neighboring atom.

It is important to stress that the reverse of the pair production is also possible. This is known as electron-positron annihilation:

$$e^- + e^+ \rightarrow \gamma + \gamma . \tag{3.70}$$

In this case the electron-positron pair vanishes and two photons appear in its place. The conservation of energy and linear momentum forbid the creation of only one photon in free space. In the most common case, the two photons have an energy equal to the rest energy $m_e c^2$ of the electron.

Further Reading

There are many books discussing the topics of this chapter. Two best-selling ones are:

Serway, R.A., Moses, C.J., Moyer, C.A.: Modern Physics. Brooks/Cole Publishing Company (2004).

Bransden, B.H., Joachain, C.J.: Physics of Atoms and Molecules. Prentice Hall (2003).

Relevant historical papers about the quantum nature of light are:

Planck, M.: Verhandlungen der Deutschen Physikalischen Gesellschaft **2**, 202 (1900).

Planck, M.: Verhandlungen der Deutschen Physikalischen Gesellschaft **2**, 237 (1900).

Einstein, A.: Annalen der Physik **17**, 132 (1905).

Compton, A.H.: Phys. Rev. **21**, 483 (1923).

Blackett, P.M.S., Occhialini, G.P.S.: Proc. R. Soc. Lond. A **139**, 699 (1933).

Chapter 4
Quantum Properties of Matter

In this chapter we analyze empirical phenomena which are explained by taking into account the quantum nature of matter. We investigate the peculiar behavior shown by the heat capacity in solids at low temperature, introducing the concept of vibrational phonons. Following the approach of Einstein and Debye, the solid can be seen as a gas of phonons characterized by a specific quantum thermal distribution, similar to the one of the photons of the black-body radiation. We also consider the quantization of the energy levels proposed in 1913 by Bohr for the hydrogen atom and the three main mechanisms of electromagnetic transitions for atoms and molecules: absorption, spontaneous emission, and stimulated emission.

4.1 Heat Capacity of Solids: Einstein Versus Debye

In 1907 Albert Einstein observed that the quantization adopted by Max Planck for the black-body radiation of photons can be used also for the harmonic vibrations of atoms in solids, which are called *phonons*. Einstein invoked a sort of generalization of the equipartition principle of the harmonic oscillator: for each mode of harmonic oscillation with frequency ν, the average thermal energy is given by

$$\bar{E}(\nu) = \frac{h\nu}{e^{\beta h\nu} - 1} , \tag{4.1}$$

that is exactly Eq. (3.14) of the previous chapter setting $\beta = 1/(k_B T)$ with k_B the Boltzmann constant and T the absolute temperature. As seen in the previous chapter, for $h\nu \ll k_B T$ one gets

$$\bar{E}(\nu) = \frac{1}{\beta} = k_B T , \tag{4.2}$$

© The Author(s), under exclusive license to Springer Nature Switzerland AG 2022
L. Salasnich, *Modern Physics*, UNITEXT for Physics,
https://doi.org/10.1007/978-3-030-93743-0_4

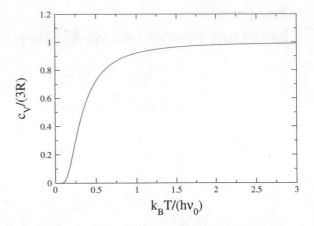

Fig. 4.1 Molar heat capacity c_V as a function of the temperature T according to the Einstein's theory of phonons in solids

that is the classical equipartition principle for the harmonic oscillator. Einstein assumed that the frequency is the same for all the $3N$ modes of a specific solid, i.e. where N is the number of atoms in the solid and the factor 3 is due to the fact that there are three possibile directions of oscillation. We denote ν_0 the Einstein's frequency of oscillation of phonons. Under these assumptions the internal energy E reads

$$E = 3\,N\bar{E}(\nu_0) = \frac{3Nh\nu_0}{e^{\beta h\nu_0} - 1} \tag{4.3}$$

and the molar heat capacity at constant volume is

$$c_V = \frac{1}{n}\left(\frac{\partial E}{\partial T}\right)_{V,N} = 3R\left(\frac{h\nu_0}{k_B T}\right)^2 \frac{e^{\frac{h\nu_0}{k_B T}}}{\left(e^{\frac{h\nu_0}{k_B T}} - 1\right)^2}, \tag{4.4}$$

where $n = N/N_A$ is the number of moles, N_A is the Avogadro number, and $R = k_B N_A$ is the gas constant. At high temperatures from this Einstein formula one recovers the Dulong-Petit law $c_V = 3R$. The measurements of the molar heat capacity performed by Walther Nernst in many solids at low temperature were in reasonable agreement with the Einstein formula, Eq. (4.4), by using ν_0 as a fitting parameter for each solid (Fig. 4.1).

In 1912 Peter Debye found that a slightly different approach is more successful in reproducing the experimental data. Debye suggested that, as in the case of the black-body radiation, the atoms in a crystal lattice can have many frequencies ν of oscillations. The total number of phonons (harmonic modes of vibrations) in a solid is given by $3N$, with N the number of atoms in the solid. As a consequence, it must be

$$3N = \int_0^{\nu_D} \mathcal{D}_1(\nu)d\nu, \tag{4.5}$$

where ν_D is the maximal admitted frequency of oscillation, which is now called Debye frequency, and $\mathcal{D}_1(\mu)$ is the single-mode density of states, i.e. the density of states of phonons. In analogy with the black-body radiation, Debye assumed that

$$\mathcal{D}_1(\nu) = A \nu^2,\qquad(4.6)$$

where the constant A is fixed by Eq. (4.5). One immediately finds

$$A = \frac{9N}{\nu_D^3}.\qquad(4.7)$$

The total internal energy of Debye is then given by

$$E = \int_0^{\nu_D} \mathcal{D}_1(\nu)\,\bar{E}(\nu)\,d\nu = \frac{9N}{\nu_D^3}\int_0^{\nu_D} \nu^2 \frac{h\nu}{e^{\frac{h\nu}{k_B T}} - 1}\qquad(4.8)$$

and the molar heat capacity at constant volume of Debye can be written as

$$c_V = 9R \left(\frac{T}{T_D}\right)^3 \int_0^{T_D/T} \frac{x^4 e^x}{(e^x - 1)^2}\,dx,\qquad(4.9)$$

where $T_D = h\nu_D/k_B$ is the Debye temperature, which depends on the specific solid material. Several crystalline solids show an extremely good agreement of the Debye equation (4.9) with experimental data. From Eq. (4.9) one finds that $c_V = 3R$ for $T/T_D \gg 1$, that is the Dulong-Petit law, while $c_V = (12R\pi^4/5)(T/T_D)^3$ for $T/T_D \ll 1$. For the sake of completeness, it is important to stress that, at very low temperature (below 1 K), one must also take into account the electronic heat capacity to reproduce correctly the experimental findings of solids.

4.2 Energy Spectra of Atoms

All the experimental data obtained with atomic gases show that individual atoms emit light only for certain characteristic wavelengths λ. Moreover, in the case of atoms, the emission wavelengths coincide with the absorption wavelengths. These wavelengths λ (or equivalently the corresponding frequencies $\nu = c/\lambda$), specific for each atom, are called *electromagnetic spectrum* of that atom. Historically, systematic experiments to determine the electromagnetic spectrum of atoms began in the mid-1800s. The technique used was to confine an atomic gas inside an ampoule (lamp), and to heat the gas by electrical methods to produce light. For example, in the discharge lamp the ends of the lamp there are two charged electrodes.

4.2.1 Energy Spectrum of Hydrogen Atom

In the case of the electromagnetic spectrum of the hydrogen atom, the numerous
experimental data are well summarized by the Rydberg empirical formula

$$\frac{1}{\lambda} = R_H \left(\frac{1}{n_1^2} - \frac{1}{n_2^2} \right) \tag{4.10}$$

where λ is one of the wavelengths of the spectrum,

$$R_H = 1.1 \cdot 10^7 \text{ m}^{-1} \tag{4.11}$$

is Rydberg's constant, while n_1 and n_2 are two natural number such that $n_2 > n_1$.
In 1888 Eq. (4.10) was obtained Johannes Rydberg as an empirical generalization
of Balmer series, previously found for the hydrogen atom. The most experimentally
studied spectral series of the hydrogen atom are: the Lyman series, with $n_1 = 1$ and
$n_2 > 1$; the Balmer series, with $n_1 = 2$ and $n_2 > 2$; the Paschen series, with $n_1 = 3$
and $n_2 > 3$; and the Brackett series, with $n_1 = 4$ and $n_2 > 4$.

Since the relation

$$\lambda \nu = c \tag{4.12}$$

is always valid, the Rydberg formula can also be written as follows

$$\nu = c\, R_H \left(\frac{1}{n_1^2} - \frac{1}{n_2^2} \right) \tag{4.13}$$

where ν is one of the frequencies of the electromagnetic spectrum of the hydrogen
atom. As we will see, Bohr explained the Rydberg empirical formula with a planetary
microscopic model of the hydrogen atom. To do this, Bohr introduced a revolutionary
hypothesis: the quantization of the motion of the electron.

4.3 Bohr's Model of Hydrogen Atom

In 1913 Niels Bohr was able to explain the discrete frequencies of electromagnetic
emission of hydrogen atom under the hypothesis that the energy of electron orbiting
around the nucleus is quantized according to formula

$$E_n = -\frac{m_e e^4}{8\varepsilon_0^2 h^2} \frac{1}{n^2} = -13.6\,\text{eV}\,\frac{1}{n^2}, \tag{4.14}$$

where $n = 1, 2, 3, \ldots$ is called *principal quantum number*, m_e is the mass of the
electron, $-e$ is the electric charge of the electron, and ε_0 is the dielectric constant in
the vacuum. Remember that

$$1 \text{ eV} = 1eV = 1.6 \cdot 10^{-19} \text{J}.$$

Equation (4.14) shows that the *quantum states* of the system are characterized by the quantum number n and the *ground state* ($n = 1$) has an energy equal to -13.6 eV, which is the ionization energy of the hydrogen atom. Adopting a notation introduced by Paul Dirac in 1939, we denote with the symbol $|n\rangle$, in words "ket n", the quantum state of the electron characterized by the principal quantum number n. According to the Bohr model, the electromagnetic radiation is emitted or absorbed only when an electron has a transition from one energy level E_{n_1} to another E_{n_2}. In addition, the frequency ν of the radiation is related to the energies of the two quantum states $|n_1\rangle$ and $|n_2\rangle$ involved in the transition according to

$$h\nu = E_{n_1} - E_{n_2}. \tag{4.15}$$

Thus, any electromagnetic transition between two quantum states involves the emission or absorption of a photon with an energy $h\nu$ equal to the energy difference of the two quantum states.

4.3.1 Derivation of Bohr's Formula

The energy of the electron with mass m_e and electric charge $q = -e$ (with $e > 0$) orbiting with uniform circular motion around the proton is given by

$$E = \frac{1}{2} m_e v^2 - \frac{e^2}{4\pi\epsilon_0 r}, \tag{4.16}$$

where the first term represents the non-relativistic kinetic energy of the electron and the second term is the potential energy due to the Coulomb electric force between the electron and the proton. As is well known, the magnitude of the electric force **F** by Coulomb is given by

$$F = \frac{e^2}{4\pi\epsilon_0 r^2}. \tag{4.17}$$

According to the second law of dynamics it must be

$$F = m_e a = m_e \frac{v^2}{r}, \tag{4.18}$$

where the last equality is obtained taking into account that in the circular motion the acceleration a is related to the velocity v and to the radius of the orbit r by $a = v^2/r$. Putting together Eqs. (4.17) and (4.18) we obtain immediately

$$\frac{1}{2}m_e v^2 = \frac{1}{2}\frac{e^2}{4\pi\epsilon_0 r} \,.$$ (4.19)

Therefore the kinetic energy of the electron is equal to half of its potential energy changed of sign. From Eq. (4.16) it follows that the energy of the electron can be written as

$$E = -\frac{1}{2}\frac{e^2}{4\pi\epsilon_0 r} \,.$$ (4.20)

The fact that there is a minus sign ensures that the electron is in a bound state. This is correct since it is assumed that the electron is bound on a circular orbit. However, this fully classical plnetary model for the electron (proposed by Hantaro Nagaoka in 1904) does not take into account that the electron, being an accelerating charged particle, emits electromagnetic radiation: losing energy the electron spirals into the proton within a fraction of a second.

Bohr observed that in this planetary model besides the energy E there is another conserved quantity: the total angular momentum $\mathbf{L}$ of the electron. The modulus of $\mathbf{L}$ is given by

$$L = |\mathbf{r} \wedge \mathbf{p}| = r\, m_e\, v\, \sin(\theta) = r\, m_e\, v$$ (4.21)

Bohr's fundamental hypothesis is that this orbital angular momentum $\mathbf{L}$ is quantized. Since the angular momentum has the units Joule $\times$ second, which are the same as Planck's constant h, Bohr in his seminal scientific paper wrote

$$L = \frac{h}{2\pi}n$$ (4.22)

where n is a natural number, i.e. $n = 1, 2, 3, 4, \dots$. From the quantization of angular momentum, explicitly

$$r\, m_e\, v = \frac{h}{2\pi}n \,,$$ (4.23)

a constraint follows between the velocity v and the radius r. On the other hand, Eq. (4.19) also gives a constraint between these two quantities. After a simple algebra with the Eqs. (4.19) and (4.23) we derive the formulas

$$r = \frac{\epsilon_0 h^2}{\pi m_e e^2}\, n^2 = r_B\, n^2$$ (4.24)

$$v = \frac{e^2}{2\epsilon_0 h}\frac{1}{n} = v_B\frac{1}{n}$$ (4.25)

where $r_B = 0.53 \cdot 10^{-10}$ m is called Bohr radius and $v_B = 2 \cdot 10^6$ m/s is the Bohr velocity. Therefore the quantization of the angular momentum implies the quantization of the radius of the orbit and also the quantization of the velocity of the electron in that orbit. The other relevant hypothesis of Bohr's model is that the electron does not emit radition until it moves in a quantized orbit. Instead, the radiation is emitted or absorbed when the electron jumps from one quantized orbit to another.

4.4 Energy Levels and Photons

Inserting Eq. (4.24) into Eq. (4.20) we obtain exactly Eq. (4.14), which can also be written as

$$E_n = -\frac{E_B}{n^2}, \tag{4.26}$$

where $E_B = (m_e e^4)/(8\varepsilon_0^2 h^2) = 13.6$ eV is the Bohr energy. As already pointed out, a single photon is emitted or absorbed when an electron has a transition from one energy level E_{n_1} to another E_{n_2}. Moreover, the energy $h\nu$ of this photon reads

$$h\nu = E_{n_2} - E_{n_1} = \frac{m_e e^4}{8\varepsilon_0^2 h^2}\left(\frac{1}{n_1^2} - \frac{1}{n_2^2}\right). \tag{4.27}$$

Recalling that

$$\nu = \frac{c}{\lambda} \tag{4.28}$$

we find

$$h\frac{c}{\lambda} = \frac{m_e e^4}{8\varepsilon_0^2 h^2}\left(\frac{1}{n_1^2} - \frac{1}{n_2^2}\right) \tag{4.29}$$

namely

$$\frac{1}{\lambda} = \frac{m_e e^4}{8c\varepsilon_0^2 h^3}\left(\frac{1}{n_1^2} - \frac{1}{n_2^2}\right). \tag{4.30}$$

This is exactly the Rydberg formula, Eq. (4.10), where

$$R_H = \frac{m_e e^4}{8c\varepsilon_0^2 h^3} \tag{4.31}$$

turns out to be the Rydberg constant, and clearly $E_B = hcR_H$.

4.5 Electromagnetic Transitions

The three main mechanisms of electromagnetic transitions in atoms, molecules, and solids are: absorption, spontaneous emission, and stimulated emission. In absorption, the electron moves from a quantum state $|a\rangle$ of lower energy E_a to quantum state $|b\rangle$ of higher energy E_b with the absorption of a photon of linear frequency $\nu =$

$\nu_{ba} = (E_b - E_a)/h$ or, equivalently of angular frequency $\omega = \omega_{ba} = (E_b - E_a)/\hbar$. In the spontaneous emission, the electron spontaneously jumps from a quantum state $|b\rangle$ of higher energy E_b to quantum state $|a\rangle$ of lower energy E_a with the emission a photon of linear frequency $\nu = \nu_{ba} = (E_b - E_a)/h$ or, equivalently of angular frequency $\omega = \omega_{ba} = (E_b - E_a)/\hbar$. In the stimulated emission, the electron is stimulated to jump from a quantum state $|b\rangle$ of higher energy E_b to quantum state $|a\rangle$ of lower energy E_a by the presence of a photon of linear frequency $\nu = \nu_{ba} = (E_b - E_a)/h$ or, equivalently of angular frequency $\omega = \omega_{ba} = (E_b - E_a)/\hbar$. During the stimulated emission it is emitted a photon with the same frequency, the same direction, and the same phase of the stimulating one.

4.6 Einstein Coefficients

In 1916 Albert Einstein observed that, given an ensemble of N atoms in two possible atomic states $|a\rangle$ and $|b\rangle$, with $N_a(t)$ the number of atoms in the state $|a\rangle$ at time t and $N_b(t)$ the number of atoms in the state $|b\rangle$ at time t, it must be

$$N = N_a(t) + N_b(t) \tag{4.32}$$

and consequently

$$\frac{dN_a}{dt} = -\frac{dN_b}{dt}. \tag{4.33}$$

Notice that, within the semiclassical approach of Bohr for the hydrogen atom, $|a\rangle$ means the electron with quantum number n_a and $|b\rangle$ means the electron with quantum number n_b. Einstein suggested that, if the atoms are exposed to an electromagnetic radiation of energy density per unit of angular frequency, $\rho(\omega)$, the rate of change of the number of atoms in the state $|a\rangle$ must be

$$\frac{dN_a}{dt} = A_{ba} N_b + B_{ba} \rho(\omega_{ba}) N_b - B_{ab} \rho(\omega_{ba}) N_a. \tag{4.34}$$

where the parameters A_{ba}, B_{ba}, and B_{ab} are known as Einstein coefficients. Here $\omega_{ba} = (E_b - E_a)/\hbar$ is the so-called Bohr angular frequency of the electromagnetic transition from the quantum state with energies E_b and E_a. The first two terms in Eq. (4.34) describe the increase of the number of atoms in $|a\rangle$ due to spontaneous and stimulated transitions from $|b\rangle$, while the third term takes into account the decrease of the number of atoms in $|a\rangle$ due to absorption with consequent transition to $|b\rangle$.

Einstein was able to determine the coefficients A_{ba}, B_{ba} and B_{ab} by supposing that the two rates in Eqs. (4.33) and (4.34) must be equal to zero at thermal equilibrium, i.e.

$$\frac{dN_a}{dt} = -\frac{dN_b}{dt} = 0, \tag{4.35}$$

In this way Einstein found

$$A_{ba}\frac{N_b}{N_a} = \rho(\omega_{ba})\left(B_{ab} - B_{ba}\frac{N_b}{N_a}\right). \tag{4.36}$$

Because the relative population of the atomic states $|a\rangle$ and $|b\rangle$ is given by a Boltzmann factor

$$\frac{N_b}{N_a} = \frac{e^{-\beta E_b}}{e^{-\beta E_a}} = e^{-\beta(E_b - E_a)} = e^{-\beta\hbar\omega_{ba}}, \tag{4.37}$$

Einstein got

$$\rho(\omega_{ba}) = \frac{A_{ba}}{B_{ab}e^{\beta\hbar\omega_{ba}} - B_{ba}}. \tag{4.38}$$

At thermal equilibrium we know that

$$\rho(\omega_{ba}) = \frac{\hbar\omega_{ba}^3}{\pi^2 c^3}\frac{1}{e^{\beta\hbar\omega_{ba}} - 1}. \tag{4.39}$$

It follows that

$$A_{ba} = B_{ba}\frac{\hbar\omega_{ba}^3}{\pi^2 c^3}, \qquad B_{ab} = B_{ba}. \tag{4.40}$$

Notice that in this way Einstein obtained the coefficient A_{ba} of spontaneous decay by calculating the coefficient of stimulated decay B_{ba}. We do not show the calculation but we report the final result for the hydrogen atom:

$$B_{ba} = \frac{\pi^2 c^3}{\hbar\omega_{ba}^3}\frac{\omega_{ba}^3}{3\pi\epsilon_0\hbar c^3}|\mathbf{d}_{ba}|^2, \tag{4.41}$$

where $\mathbf{d}_{ba} = e\mathbf{r}_{ba}$ is the so-called electric dipole element with e the charge of the proton and $\mathbf{r}_{ba}$ the position vector associated to the distance between the quantum states $|a\rangle$ and $|b\rangle$. In the case of the semiclassical model of Bohr we have $|\mathbf{r}_{ba}| = |r_{n_b} - r_{n_a}|$, where r_n is radius of the Bohr stationary orbit of the electron with quantum number n.

It is important to stress that the laser device, invented in 1957 by Charles Townes and Arthur Schawlow at Bell Labs, is based on a generalization of Eqs. (4.33) and (4.34) describing the light amplification by stimulated emission of radiation.

4.7 Life-Time of an Atomic State

We have seen that the Einstein coefficient A_{ba} gives the transition probability per unit time from the atomic state $|b\rangle$ to the atomic state $|a\rangle$. This means that, according to Einstein, in the absence of an external electromagnetic radiation one has

$$\frac{dN_b}{dt} = -A_{ba}\, N_b \qquad (4.42)$$

with the unique solution

$$N_b(t) = N_b(0)\, e^{-A_{ba}t} \,. \qquad (4.43)$$

It is then quite natural to consider $1/A_{ba}$ as the characteristic time of this spontaneous transition. More generally, the life-time τ_b of an atomic state $|b\rangle$ is defined as the reciprocal of the total spontaneous transition probability per unit time to all possible final atomic states $|a\rangle$, namely

$$\tau_b = \frac{1}{\sum_a A_{ba}} \,. \qquad (4.44)$$

Clearly, if $|b\rangle$ is the ground-state then $A_{ba} = 0$ and $\tau_b = \infty$.

Let us perform an instructive exercise. 10^8 sodium atoms are excited to the first excited state of sodium by absorption of light. Knowing that the excitation energy is 2.125 eV and the lifetime is 16 ns, we want to calculate the maximum of the emitting power. The total absorbed energy is given by

$$E = N\epsilon \,, \qquad (4.45)$$

where N is the number of atoms (and also the number of absorbed photons) while ϵ is the transition energy

$$\epsilon = 2.125 \,\text{eV} = 2.125 \cdot 1.6 \cdot 10^{-19} \,\text{J} = 3.4 \cdot 10^{-19} \,\text{J} \,. \qquad (4.46)$$

The absorbed energy is then

$$E = 10^8 \cdot 3.4 \cdot 10^{-19} \,\text{J} = 3.4 \cdot 10^{-11} \,\text{J} \,. \qquad (4.47)$$

The absorbed energy is equal to the energy emitted by spontaneous de-excitation. The emitting power $P(t)$ decays exponentially with time t as

$$P(t) = P_0 e^{-t/\tau} \,, \qquad (4.48)$$

where P_0 is the maximum of the emitting power and $\tau = 16\,\text{ns}$ is the lifetime of the excited state. It must be

$$E = \int_0^\infty P(t)\, dt = P_0\,\tau \,, \qquad (4.49)$$

from which we get

$$P_0 = \frac{E}{\tau} = \frac{3.4 \cdot 10^{-11} \,\text{J}}{16 \cdot 10^{-9} \,\text{s}} = 2.1 \cdot 10^{-3} \,\text{J/s} = 2.1 \cdot 10^{-3} \,\text{W} \,. \qquad (4.50)$$

4.8 Natural Line Width

Many experimental results show that, in the radiation energy spectrum, the natural line-width Γ_N due to the transition from the state $|b\rangle$ to the state $|a\rangle$ is accurately described by the formula

$$\Gamma_N = \hbar \left(\frac{1}{\tau_b} + \frac{1}{\tau_a} \right) . \tag{4.51}$$

Moreover, in this transition the intensity of the emitted electromagnetic radiation follows the Lorentzian peak

$$I(\epsilon) = \frac{I_0 \Gamma_N^2 / 4}{(\epsilon - E_{ba})^2 + \Gamma_N^2 / 4}, \tag{4.52}$$

where $\epsilon = \hbar \omega$ is the energy of the emitted photon and $E_{ba} = E_b - E_a$ is the energy difference of the two atomic states. The Lorentzian peak is centered on $\epsilon = E_{ba}$ and Γ_N is its full width at half-maximum. These empirical results are not explained by the old semiclassical quantum mechanics of Bohr, but are instead quite well explained by the modern quantum mechanics we shall describe in next chapters.

As an example, we want to calculate the natural frequency width of the α-Lymann line in a hydrogen gas at the temperature $1000\,\mathrm{K}$, knowing that the lifetime of the excited state is $0.16 \cdot 10^{-8}$ s and the wavelength of the transition is $1214 \cdot 10^{-10}$ m. The natural width between the states $|i\rangle$ and $|j\rangle$ is given by

$$\Delta \nu_N = \frac{1}{2\pi} \left(\frac{1}{\tau_i} + \frac{1}{\tau_j} \right) \tag{4.53}$$

where τ_i is the lifetime of the state $|i\rangle$ and τ_j is the lifetime of the state $|j\rangle$. In our problem $\tau_{1s} = \infty$, because $|1s\rangle$ is the ground-state, while $\tau_{2p} = 0.16 \cdot 10^{-8}$ s for the excited state $|2p\rangle$. It follows that

$$\Delta \nu_N = \frac{1}{2\pi} \left(\frac{1}{\tau_{1s}} + \frac{1}{\tau_{1p}} \right) = \frac{1}{6.28} \left(0 + \frac{1}{0.16 \cdot 10^{-8}} \right) \, \mathrm{s}^{-1} = 9.9 \cdot 10^7 \, \mathrm{Hz} . \tag{4.54}$$

4.8.1 Collisional Broadening

It is important to observe that the effective line-width Γ measured in the experiments is usually larger than Γ_N because the radiating atoms move and collide. In fact, one can write

$$\Gamma = \Gamma_N + \Gamma_C + \Gamma_D , \tag{4.55}$$

where in addition to the natural width Γ_N there are the so-called collisional broadening width Γ_C and the Doppler broadening width Γ_D.

The collisional broadening is due to the collisions among the identical atoms of a gas. The collision reduces the effective life-time of an atomic state and the collisional width can be then written as

$$\Gamma_C = \frac{\hbar}{\tau_{col}}, \tag{4.56}$$

where τ_{col} is the collisional time, i.e. the average time between two collision of atoms in the gas. According to the results of statistical mechanics, τ_{col} is given by

$$\tau_{col} = \frac{1}{n\sigma v_{mp}}, \tag{4.57}$$

where n is the number density of atoms, σ is the cross-section, and v_{mp} is the most probable speed of the particles in the gas. The cross-section σ, which can be interpreted as the effective area seen by a particle colliding with another particle, is clearly dependent on the inter-atomic potential energy between two atoms. Moreover, by considering the Maxwell-Boltzmann distribution of speeds in an ideal gas, the most probable speed v_{mp} is given by

$$v_{mp} = \sqrt{\frac{2k_B T}{m}}, \tag{4.58}$$

where T is the absolute temperature, k_B is the Boltzmann constant and m is the mass of each particle.

As an example, let us calculate the collisional frequency width of the α-Lyman line for a gas of hydrogen atoms with density 10^{12} atoms/(m^3) and collisional cross-section 10^{-19} m^2 at the temperature 10^3 K. The α-Lyman line is associated to the $1s \rightarrow 2p$ transition of the hydrogen atom. The collisional frequency width is given by

$$\Delta\nu_C = \frac{1}{2\pi\tau_{col}}, \tag{4.59}$$

where τ_{col} is the collision time. This time depends on the cross-section $\sigma = 10^{-19}$ m^2 and on the gas density $n = 10^{12}$ m^{-3} according to the formula

$$\tau_{col} = \frac{1}{n\ \sigma\ v_{mp}}, \tag{4.60}$$

where $v_{mp} = \sqrt{2k_B T/m_H}$ is the velocity corresponding the the maximum of the Maxwell-Boltzmann distribution. We have then

$$\Delta\nu_C = \frac{n\ \sigma}{2\pi}\sqrt{\frac{2k_B T}{m_H}}. \tag{4.61}$$

Because $T = 10^3$ K, $k_B = 1.3 \cdot 10^{-23}$ J/K and $m_H = 1.6 \cdot 10^{-27}$ kg, we finally obtain

$$\Delta \nu_C = 6.4 \cdot 10^{-5} \text{ Hz .} \tag{4.62}$$

4.8.2 Doppler Broadening

The Doppler broadening is due to the Doppler effect caused by the distribution of velocities of atoms. For non-relativistic velocities ($v_x \ll c$) the Doppler shift in frequency is

$$\omega = \omega_0 \left(1 - \frac{v_x}{c} \right) , \tag{4.63}$$

where ω is the observed angular frequency, ω_0 is the rest angular frequency, v_x is the component of the atom speed along the axis between the observer and the atom and c is the speed of light. The Maxwell-Boltzmann distribution $f(v_x)$ of speeds $v_x = -c(\omega - \omega_0)/\omega_0$ at temperature T, given by

$$f(v_x)\, dv_x = \left(\frac{m}{2\pi k_B T} \right)^{1/2} e^{-m v_x^2 / (2k_B T)}\, dv_x , \tag{4.64}$$

becomes

$$f(\omega)\, d\omega = \left(\frac{m}{2\pi k_B T} \right)^{1/2} e^{-m c^2 (\omega - \omega_0)^2 / (2\omega_0^2 k_B T)} \frac{c}{\omega_0}\, d\omega \tag{4.65}$$

in terms of the angular frequency ω. This is the distribution of frequencies seen by the observer, and the full width at half-maximum of the Gaussian is taken as Doppler width, namely

$$\Gamma_D = \sqrt{\frac{8 \ln (2) k_B T}{mc^2}}\, \hbar \omega_0 . \tag{4.66}$$

Also in this case we perform a simple exercise. We want to calculate the Doppler frequency width of the α-Lyman line in a hydrogen gas at the temperature 1000 K, knowing that the lifetime of the excited state is $0.16 \cdot 10^{-8}$ s and the wavelength of the transition is $1214 \cdot 10^{-10}$ m. The Doppler width depends on the temperature T and on the frequency ν of the transition according to the formula

$$\Delta \nu_D = \nu \sqrt{8 \ln(2)} \sqrt{\frac{k_B T}{m_H c^2}} = \frac{\nu}{c} \sqrt{8 \ln(2)} \sqrt{\frac{k_B T}{m_H}} , \tag{4.67}$$

where $k_B = 1.3 \cdot 10^{-23}$ J/K is the Boltzmann constant, $m_H = 1.6 \cdot 10^{-27}$ kg is the mass of an hydrogen atom, while $c = 3 \cdot 10^8$ m/s is the speed of light in the vacuum. We know that

$$\frac{\nu}{c} = \frac{1}{\lambda} = \frac{1}{1.216 \cdot 10^{-7}\,m} = 8.22 \cdot 10^6\,\text{m}^{-1}\,, \tag{4.68}$$

and then we get

$$\Delta \nu_D = 8.22 \cdot 10^6 \cdot 2.35 \cdot \sqrt{\frac{1.3 \cdot 10^{-23} \cdot 10^3}{1.6 \cdot 10^{-27}}}\,\text{Hz} = 5.6 \cdot 10^{10}\,\text{Hz}\,. \tag{4.69}$$

4.9   Old Quantum Mechanics of Bohr, Wilson and Sommerfeld

The early quantum results of Planck, Einstein, Bohr and others are substantially characterized by heuristic prescriptions. Such results, far from being able to be considered complete, have had the merit to anticipate, at least in the spirit, the modern quantum mechanics. In the case of matter with mass different from zero a quite general extension of Bohr 1913 results for the hydrogen atom was proposed in 1915-1916 by William Wilson and Arnold Sommerfeld. This Bohr-Wilson-Sommerfeld theory, known as "old quantum mechanics" allows that the orbits of the electron in the hydrogen atom may be elliptical and not just circular, as assumed by Niels Bohr.

The premise of the old quantum theory is that the classical Hamiltonian $H(\vec{q},\,\vec{p})$ of a three-dimensional system is separable, i.e. $H(\vec{q},\,\vec{p}) = H_1(q_1,\,p_1) + H_2(q_2,\,p_2) + H_3(q_3,\,p_3)$, where q_i are the (generalized) coordinates and p_i are the corresponding linear momenta. Then, not all motions are permitted, but only those that comply with the quantization condition

$$\oint_{H(\mathbf{q},\vec{p})=E} p_i\,dq_i = n_i\,h\,, \tag{4.70}$$

where the quantum numbers n_i are integers, and the integral is taken over one period of the classical motion at constant energy E.

Unfortunately, this old quantum theory works reasonably well only if the system under investigation is integrable, i.e. if there is a canonical transformation of coordinates which makes the Hamiltonian separable. Moreover, this theory does not take into account the dual wave-particle nature of the electron.

Further Reading

The quantum mechanical approach of Einstein and Debye to the heat capacity of solids is very well described in the book:
J.D. McGervey, Introduction to Modern Physics (Academic Press, 1983).
Relevant historical papers about the Rydberg formula, the Bohr model for the hydro-

gen atom, and the Einstein coefficients of electromagnetic transitions are:

J.R. Rydberg, Proceedings of the Royal Swedish Academy of Science, **23**, 1 (1889).

N. Bohr, Philosophical Magazine **26**, 1 (1913).

A. Einstein, Verhandlungen der Deutschen Physikalischen Gesellschaft **18**, 318 (1916).

For a quite complete discussion of line widths in elecromagnetic spectra:

B.H. Bransden and C.J. Joachain, Physics of Atoms and Molecules (Prentice Hall, 2003).

The derivation of the old-quantum mechanics of Bohr-Wilson-Sommerfeld from the more modern Schrödinger equation is discussed in several papers. See, for instance,

M. Robnik and L. Salasnich, Journal of Physics A **30**, 1719 (1997).

A detailed calculation of the Einstein coefficient of spontaneous emission can be found in the book:

L. Salasnich, Quantum Physics of Light and Matter. Photons, Atoms, and Strongly Correlated Systems (Springer, 2019).

Chapter 5
Wavefunction of a Quantum Particle

In this chapter we introduce the time-dependent Schrödinger equation, obtained in 1926 by Erwin Schrödinger from the revolutionary idea of De Broglie to associate to each particle, and in particular to the electron, a quantum wave. We also discuss the Born probabilistic interpretation of the Schrödinger wavefunction, which is the interpretation commonly accepted even today because it has been confirmed by several sophisticated experiments.

5.1 De Broglie Wavelength

Inspired by the behavior of light, that shows both wave and corpuscular properties, in the period 1922–1924 Louis de Broglie suggested that also the electron has wave proprties. De Broglie postulated that the relation

$$\lambda = \frac{h}{p} \tag{5.1}$$

applies not only to photons but also to matter particles. Thus, this equation applies also to the electron and, more generally, to particles with mass m different from zero. Remember that p is the magnitude of linear momentum (corpuscular property) and λ is the wavelength (wave property) of the "quantum particle". The quantum particle is therefore characterized both by wave and corpuscular properties. This is the so-called *wave-particle duality* introduced by De Broglie, and then strongly supported by Niels Bohr, Werner Heisenberg and others (Copenhagen interpretation of quantum mechanics).

L. Salasnich, *Modern Physics*, UNITEXT for Physics,
https://doi.org/10.1007/978-3-030-93743-0_5

5.1.1   Explaining the Bohr Quantization

On the basis of his hypothesis, De Broglie was able to explain the quantization proposed by Bohr for orbital angular momentum of the electron in the hydrogen atom. De Broglie's reasoning is the following:

(i) the electron of mass m_e, velocity v and momentum $p = m_e v$ is also a wave;

(ii) this electronic matter wave is characterized by the wavelength $\lambda = h/p = h/(m_e v)$;

(iii) in a closed trajectory of length l the electronic wave the electron wave is stable only if

$$l = \lambda n , \tag{5.2}$$

where n is a natural number. Equation (5.2) is well known in wave motion theory and ensures that there are no distructive interference effects in a wave. A wave which satisfies this equation is called stationary wave. In the case of circular motion of radius r we will have

$$l = 2\pi r \tag{5.3}$$

and moreover for the electron is

$$\lambda = \frac{h}{m_e v} . \tag{5.4}$$

So the formula (5.2) becomes

$$2\pi r = \frac{h}{m_e v} n \tag{5.5}$$

i.e.

$$r \, m_e v = \frac{h}{2\pi} n . \tag{5.6}$$

This is precisely the quantization formula of the angular momentum $L = r m_e v$ of Bohr:

$$L = \hbar n \tag{5.7}$$

with $\hbar = h/(2\pi)$.

5.2   Wave Mechanics of Schrödinger

In 1926 Erwin Schrödinger introduced the equation that bears his name. According to Schrödinger, a particle of mass m under the action of a conservative external force $\mathbf{F}$, i.e. such that

$$\mathbf{F} = -\nabla U(\mathbf{r}) , \tag{5.8}$$

where $U(\mathbf{r})$ is the potential energy of the particle in the position $\mathbf{r} = (x, y, z)$, is characterized by a wave function $\psi(\mathbf{r}, t)$, which depends on spatial position $\mathbf{r}$ and time coordinate t. This wavefunction $\psi(\mathbf{r}, t)$ satisfies the equation

$$i\hbar \frac{\partial}{\partial t}\psi(\mathbf{r}, t) = -\frac{\hbar^2}{2m}\nabla^2\psi(\mathbf{r}, t) + U(\mathbf{r})\psi(\mathbf{r}, t) \tag{5.9}$$

known as *time-dependent Schrödinger equation*. In this equation $i = \sqrt{-1}$ is the imaginary unit of the complex numbers while $\hbar$ is the reduced Planck constant.

It is important to point out that, initially, Schrödinger thought that the complex wave function $\psi(\mathbf{r}, t)$ was a matter wave, such that $|\psi(\mathbf{r}, t)|^2$ represents the local density of electrons that are at position $\mathbf{r}$ and time t. It was Max Born in 1926 to suggest a probabilistic interpretation of the wavefunction, which is commonly accepted even today: the wavefunction $\psi(\mathbf{r}, t)$ should be understood as the complex probability amplitude, where $|\psi(\mathbf{r}, t)|^2$ gives the local probability density of finding an electron at position $\mathbf{r}$ and time t, with the normalization condition

$$\int_{\mathbb{R}^3} |\psi(\mathbf{r}, t)|^2 \, d^3\mathbf{r} = 1 \tag{5.10}$$

at any time t.

In the case of N particles the probabilistic interpretation of Born becomes crucial: $\Psi(\mathbf{r}_1, \mathbf{r}_2, ..., \mathbf{r}_N, t)$ is the many-body complex wavefunction of the systems, such that $|\Psi(\mathbf{r}_1, \mathbf{r}_2, ..., \mathbf{r}_N, t)|^2$ is the probability density of finding, at time t, a particle at position $\mathbf{r}_1$, another particle at position $\mathbf{r}_2$, and so on.

Niels Bohr (in 1922), Werner Heisenberg (in 1932), Erwin Schrödinger (in 1933) and Max Born (in 1954) were awarded the Nobel Prize in Physics for their contributions to the formulation of quantum mechanics. In particular, Max Born received the Nobel Prize "for his fundamental research in quantum mechanics, in particular for his statistical interpretation of the wave function".

5.2.1   Derivation of Schrödinger's Equation

We have seen several times the dispersion relation for monochromatic light:

$$\omega = ck, \tag{5.11}$$

where ω is the angular frequency, $\mathbf{k}$ is the wave vector, while $k = |\mathbf{k}|$ the wave number. Multiplying this relation by the reduced Planck constant $\hbar$ we get

$$\hbar\omega = c\,\hbar k, \tag{5.12}$$

namely

$$E = c\,p\,, \tag{5.13}$$

where $E = \hbar\omega = h\nu$ is the energy of a particle of light (the photon) and $p = \hbar k = h/\lambda$ is the modulus of its momentum $\mathbf{p}$, with h the unreduced Planck constant, $\nu = \omega/(2\pi)$ the linear frequency, and $\lambda = 2\pi/k$ the wavelength.

The energy E of a non-relativistic particle of mass m not subject to external forces is given by

$$E = \frac{p^2}{2m}\,, \tag{5.14}$$

where $\mathbf{p}$ is the momentum of the particle and $p = |\mathbf{p}|$ is its modulus. Following the idea of Louis de Broglie, we assume that we can write

$$E = \hbar\omega \tag{5.15}$$

$$\mathbf{p} = \hbar\mathbf{k} \tag{5.16}$$

where ω and $\mathbf{k}$ are respectively the angular frequency and the wavevector of the monochromatic quantum wave associated to the particle. It immediately follows that the dispersion relation is

$$\omega = \frac{\hbar k^2}{2m}\,. \tag{5.17}$$

It is important to note that, unlike the case of the light wave, in the quantum matter wave the dispersion relation contains explicitly on the reduced Planck constant $\hbar$.

Previously we have shown that the dispersion relation (5.11) of monochromatic light is easily obtained starting from the d'Alembert equation of the electric field

$$\left(\frac{1}{c^2}\frac{\partial^2}{\partial t^2} - \nabla^2\right)\mathbf{E}(\mathbf{r}, t) = 0\,, \tag{5.18}$$

under the assumption that the electric field is a plane wave

$$\mathbf{E}(\mathbf{r}, t) = \mathbf{E}_0\, e^{i(\mathbf{k}\cdot\mathbf{r}-\omega t)}\,. \tag{5.19}$$

The question that Erwin Schrödinger asked himself was: Given the dispersion relation (5.17) of de Broglie's quantum wave of matter, assuming that this wave is describable as a monochromatic wave of the type

$$\psi(\mathbf{r}, t) = \psi_0\, e^{i(\mathbf{k}\cdot\mathbf{r}-\omega t)}\,, \tag{5.20}$$

what is the differential equation that allows me to obtain the dispersion relation above? In 1926 Schrödinger found an answer to this question. The differential equation he searched is

$$ i\hbar \frac{\partial}{\partial t} \psi(\mathbf{r}, t) = -\frac{\hbar^2}{2m} \nabla^2 \psi(\mathbf{r}, t). \tag{5.21} $$

It is in fact easy to verify that by substituting Eq. (5.20) in Eq. (5.21) we find Eq. (5.17).

5.3 Double-Slit Experiment with Electrons

Historically, the experiment that demonstrated the wave-like nature of light was performed by Thomas Young in 1801, and is known as the Young's double-slit experiment. A single source of monochromatic light of wavelength λ. Illuminates an opaque screen with two small holes placed at distance d. The slits become two sources of coherent light that generate on a screen placed at distance L an interference pattern consisting of alternating dark and bright bands. At a very large distance from the slits ($L \gg d$), the lines joining the slits with a certain point P on the screen are approximately parallel, and form an angle θ with the normal of the line joining the slits. Experimentally, the light maxima on the far panel are obtained when the angle θ satisfies the relation

$$ n\lambda = d \sin(\theta), \tag{5.22} $$

where $n = 0, 1, 2, 3, \dots$ is a natural number. This result is explained on the basis of constructive interference between light beams, which have an optical path difference $\Delta = d \sin(\theta)$. It is important to stress that the ratio λ/d is a crucial parameter in the double-slit experiment. The distance between consecutive interference fringes will be minimal if $\lambda/d \ll 1$ and the interference effects may not be seen.

The double-slit experiment was first performed using electrons in 1961 by Claus Jonsson. The experimental results fully confirmed that the beam of electrons produces an interference pattern which satisfies the formula (5.22) with the wavelength λ such that

$$ \lambda = \frac{h}{m_e v}, \tag{5.23} $$

where h is Planck's constant, m_e is the mass of the electron and v is the electron velocity of the incident beam. The experiment was repeated in 1974 in Bologna by Pier Giorgio Merli, Gianfranco Missiroli and Giulio Pozzi, who, however, sent one electron at a time on the photographic plate. The result of the Bologna experiment of 1974 is amazing:

(a) although the electrons are sent one at a time precise light and dark zones are formed on the photographic plate photographic plate;
(b) the single electron hits a point on the photographic plate photographic plate;
(c) the interference bangs are obtained as the sum of single electron events.

The double-slit experiment with electrons was also performed in 1989 by Akira Tonomura and collaborators at the Hitachi Lab by using a better resolution. Their results confirm the findigs of Merli, Missiroli and Pozzi. The single electron is therefore both a particle and a wave, as found for light photons. It should be noted that today it is possible to perform the Young's double-slit experiment also with single photons: the results are in perfect agreement with what we observed with single electrons.

5.4 Formal Quantization Rules

Starting from the monochromatic plane wave, for both the d'Alembert equation and the Schrödinger equation, the dispersion relation is obtained by taking into account that

$$\frac{\partial}{\partial t} \longleftrightarrow -i\omega \,, \tag{5.24}$$

$$\nabla \longleftrightarrow i\mathbf{k} \,, \tag{5.25}$$

i.e.: applying the time prime derivative to the plane wave function we obtain the same function multiplied by $-i\omega$, while applying the spatial prime derivative to the plane wave function one obtains the same function multiplied by $i\mathbf{k}$. Recall that $i = \sqrt{-1}$. These expressions can also be written as

$$\omega \longleftrightarrow i\frac{\partial}{\partial t} \,, \tag{5.26}$$

$$\mathbf{k} \longleftrightarrow -i\nabla \,, \tag{5.27}$$

multiplying by i and taking into account that $i^2 = -1$. Multiplying also by $\hbar$ we have then

$$\hbar\omega \longleftrightarrow i\hbar\frac{\partial}{\partial t} \,, \tag{5.28}$$

$$\hbar\mathbf{k} \longleftrightarrow -i\hbar\nabla \,. \tag{5.29}$$

Recalling that $E = \hbar\omega$ and that $\mathbf{p} = \hbar\mathbf{k}$ ultimately we get

$$E \longleftrightarrow i\hbar\frac{\partial}{\partial t} \,, \tag{5.30}$$

$$\mathbf{p} \longleftrightarrow -i\hbar\nabla \,. \tag{5.31}$$

These expressions are known as quantization rules: from the formula that links the energy E to the momentum $\mathbf{p}$ of a particle with these rules we obtain the differential equation that describes the wave associated with the particle.

5.4.1 Schrödinger Equation for a Free Particle

The energy E of a non-relativistic free particle of mass m is

$$E = \frac{p^2}{2m}, \tag{5.32}$$

where $\mathbf{p}$ is the linear momentum of the particle. Using the quantization rules (5.30) and (5.31) we obtain the Schrödinger equation of a quantum free particle

$$i\hbar \frac{\partial}{\partial t} \psi(\mathbf{r}, t) = -\frac{\hbar^2}{2m} \nabla^2 \psi(\mathbf{r}, t), \tag{5.33}$$

where $\psi(\mathbf{r}, t)$ is the matter wave function associated with the particle, usually simply called wavefunction.

5.4.2 Schrödinger Equation for a Particle in an External Potential

The energy E of a non-relativistic particle of mass m subjected to an external force of potential energy $U(\mathbf{r})$ is given by

$$E = \frac{p^2}{2m} + U(\mathbf{r}). \tag{5.34}$$

Using also in this case the rules of quantization (5.30) and (5.31) we obtain the Schrödinger equation of a quantum particle in presence of an external potential

$$i\hbar \frac{\partial}{\partial t} \psi(\mathbf{r}, t) = -\frac{\hbar^2}{2m} \nabla^2 \psi(\mathbf{r}, t) + U(\mathbf{r})\psi(\mathbf{r}, t), \tag{5.35}$$

where $\psi(\mathbf{r}, t)$ is the wavefunction associated to the particle. Clearly, this equation can be rewritten as

$$i\hbar \frac{\partial}{\partial t} \psi(\mathbf{r}, t) = \left(-\frac{\hbar^2}{2m} \nabla^2 + U(\mathbf{r}) \right) \psi(\mathbf{r}, t), \tag{5.36}$$

or, equivalently, as

$$\left(i\hbar \frac{\partial}{\partial t} + \frac{\hbar^2}{2m} \nabla^2 - U(\mathbf{r}) \right) \psi(\mathbf{r}, t) = 0. \tag{5.37}$$

Recall the probabilistic interpretation of the wave function (Max Born, 1927): $\psi(\mathbf{r}, t)$, that is also called complex *probability amplitude*, is such that its square modulus $|\psi(\mathbf{r}, t)|^2$ represents the real *probability density* of finding the particle at position $\mathbf{r}$ and time t. The probability density is sometimes written as

$$\rho(\mathbf{r}, t) = |\psi(\mathbf{r}, t)|^2 . \tag{5.38}$$

Its integral over the entire three-dimensional space $\mathbb{R}^3$ must give 1, that is

$$\int_{\mathbb{R}^3} \rho(\mathbf{r}, t) \, d^3\mathbf{r} = 1 \tag{5.39}$$

at any time t. Instead, the integral over a finite region of space V represents the probability $P_V(t)$ of finding the particle within that region V at time t. In symbols

$$\int_V \rho(\mathbf{r}, t) \, d^3\mathbf{r} = P_V(t) , \tag{5.40}$$

where, of course, $0 \le P_V(t) \le 1$ for any time t. In general, $P_V(t)$ is a function of time t.

5.5   Madelung Transformation

In 1926 Erwin Madelung found that the just discovered time-dependent Schrödinger equation can be rewritten as the Euler equations of irrotational and inviscid hydrodynamics, with a very peculiar equation of state, that is the so-called called quantum potential. In fact, setting

$$\psi(\mathbf{r}, t) = \rho(\mathbf{r}, t)^{1/2} \, e^{i\theta(\mathbf{r}, t)} , \tag{5.41}$$

and inserting this formula into Eq. (5.35) one finds

$$\frac{\partial}{\partial t}\rho + \nabla \cdot (\rho \, \mathbf{v}) = 0 , \tag{5.42}$$

$$m\frac{\partial}{\partial t}\mathbf{v} + \nabla \left[\frac{1}{2}mv^2 + U(\mathbf{r}) - \frac{\hbar^2}{2m}\frac{\nabla^2 \sqrt{\rho}}{\sqrt{\rho}} \right] = \mathbf{0} , \tag{5.43}$$

where $\rho(\mathbf{r}, t)$ is the local probability density and

$$\mathbf{v}(\mathbf{r}, t) = \frac{\hbar}{m}\nabla\theta(\mathbf{r}, t) \tag{5.44}$$

is a local velocity, that is (by definition) irrotational, i.e.

$$\nabla \wedge \mathbf{v}(\mathbf{r}, t) = \mathbf{0} . \tag{5.45}$$

Equation (5.42) is the continuity equation, with

$$\mathbf{j}(\mathbf{r}, t) = \rho(\mathbf{r}, t) \, \mathbf{v}(\mathbf{r}, t) = \frac{\hbar}{2mi} \left[\psi^*(\mathbf{r}, t) \nabla \psi(\mathbf{r}, t) - \psi(\mathbf{r}, t) \nabla \psi^*(\mathbf{r}, t) \right] \tag{5.46}$$

the current density. Integrating Eq. (5.42) over a volume V one finds

$$\frac{d}{dt} \int_V \rho \, d^3\mathbf{r} = -\int_V \nabla \cdot \mathbf{j}(\mathbf{r}, t) = -\oint_S \mathbf{j}(\mathbf{r}, t) \cdot \mathbf{n}(\mathbf{r}) , \tag{5.47}$$

taking into account the divergence theorem with S the closed surface of the volume V. In many applications, with a sufficiently large volume V, i.e. when $V = \mathbb{R}^3$, on the surface S the current density $\mathbf{j}(\mathbf{r}, t)$ is zero. Under this condition we get

$$\int_{\mathbb{R}^3} \rho(\mathbf{r}, t) \, d^3\mathbf{r} = \text{constant} , \tag{5.48}$$

which ensures the conservation of the probability, and we can set constant $= 1$.

Equation (5.43) is reminiscent of Newton's second law of dynamics. It contains a term which depends explicitly on the reduced Planck constant $\hbar$, i.e. the quantum potential

$$U_Q(\mathbf{r}, t) = -\frac{\hbar^2}{2m} \frac{\nabla^2 \sqrt{\rho(\mathbf{r}, t)}}{\sqrt{\rho(\mathbf{r}, t)}} . \tag{5.49}$$

Clearly, this is not an external potential, because it depends on the local probability density $\rho(\mathbf{r}, t)$ of the quantum fluid.

Within the Euler hydrodynamics of the Schrödinger equation, quantum effects are encoded not only in the quantum potential, but also into the properties of the local field $\mathbf{v}(\mathbf{r}, t)$. This velocity field is proportional to the gradient of a scalar field, $\theta(\mathbf{r}, t)$, that is the angle of the phase of the single-valued complex wavefunction $\psi(\mathbf{r}, t)$. Consequently, $\mathbf{v}(\mathbf{r}, t)$ satisfies the equation

$$\oint_C \mathbf{v} \cdot d\mathbf{r} = \frac{\hbar}{m} \oint_C \nabla\theta \cdot d\mathbf{r} = \frac{\hbar}{m} \oint_C d\theta = \frac{\hbar}{m} 2\pi n = \frac{h}{m} n \tag{5.50}$$

for any closed contour C, with n an integer number. In other words, the circulation is quantized in units of h/m.

5.6 Stationary Schrödinger Equation

It is important to stress that the time-dependent Schrödinger equation can be written in a more compact form as

$$i\hbar\frac{\partial}{\partial t}\psi(\mathbf{r}, t) = \hat{H}\,\psi(\mathbf{r}, t)\,, \tag{5.51}$$

where

$$\hat{H} = -\frac{\hbar^2}{2m}\nabla^2 + U(\mathbf{r}) \tag{5.52}$$

is the so-called quantum *Hamiltonian operator*, i.e. the quantum energy operator, associated to the classical total energy $E = p^2/(2m) + U(\mathbf{r})$.

Given the time-dependent Schrödinger equation (5.35), setting

$$\psi(\mathbf{r}, t) = \phi(\mathbf{r})\,e^{-iEt/\hbar} \tag{5.53}$$

one immediately finds

$$E\,\phi(\mathbf{r}) = \left(-\frac{\hbar^2}{2m}\nabla^2 + U(\mathbf{r})\right)\phi(\mathbf{r})\,, \tag{5.54}$$

known as the time-independent Schrödinger equation or *stationary Schrödinger equation*. Obviously this equation can be formally rewritten as

$$\hat{H}\phi(\mathbf{r}) = E\,\phi(\mathbf{r})\,, \tag{5.55}$$

i.e. as an eigenvalue equation. In general there are many real values of E, called *eigenvalues* (or energy levels) that satisfy this equation and the corresponding functions $\phi(\mathbf{r})$, called *eigenfunctions*. For this reason, the previous equation is usually rewritten as

$$\hat{H}\phi_{\mathbf{n}}(\mathbf{r}) = E_{\mathbf{n}}\,\phi_{\mathbf{n}}(\mathbf{r})\,, \tag{5.56}$$

where $E_{\mathbf{n}}$ is one element of the set of possible eigenvalues labelled by the subindex $\mathbf{n} = (n_1, n_2, n_3)$, which represents the possible quantum numbers. Moreover, $\phi_{\mathbf{n}}(\mathbf{r})$ is one element of the corresponding set of eigenfunctions.

Recall that to obtain the stationary Schrödinger equation we have made the hypothesis that

$$\psi(\mathbf{r}, t) = \phi(\mathbf{r})\,e^{-iEt/\hbar} \tag{5.57}$$

from which it immediately follows that

$$|\psi(\mathbf{r}, t)|^2 = |\phi(\mathbf{r})|^2 \tag{5.58}$$

provided that E is a real number. So, the wave functions satisfying Eq. (5.57) are a very particular class of solutions of the time-dependent Schrödinger equation from time: they are those for which the probability density does not depend on time. For this reason they are called stationary wavefunctions.

5.6.1 Properties of the Hamiltonian Operator

Let us consider the Hamiltonian operator $\hat{H}$ given by Eq. (5.52). It is called *hermitian* because $\hat{H}^* = \hat{H}$. It is also called *symmetric* because

$$\int_{\mathbb{R}^3} \phi_1(\mathbf{r})\hat{H}\phi_2(\mathbf{r})\, d^3\mathbf{r} = \int_{\mathbb{R}^3} \phi_2(\mathbf{r})\hat{H}\phi_1(\mathbf{r})\, d^3\mathbf{r} \tag{5.59}$$

for any choice of functions $\phi_1(\mathbf{r})$ and $\phi_2(\mathbf{r})$ which go to zero at infinity. In fact,

$$\int_{\mathbb{R}^3} \left(\phi_1(\mathbf{r})\hat{H}\phi_2(\mathbf{r}) - \phi_2(\mathbf{r})\hat{H}\phi_1(\mathbf{r})\right) d^3\mathbf{r} = \frac{\hbar^2}{2m}\int_{\mathbb{R}^3}\left(-\phi_1(\mathbf{r})\nabla^2\phi_2(\mathbf{r}) + \phi_2(\mathbf{r})\nabla^2\phi_1(\mathbf{r})\right) d^3\mathbf{r}$$

$$= \frac{\hbar^2}{2m}\int_{\mathbb{R}^3}\left(\nabla\phi_1(\mathbf{r})\cdot\nabla\phi_2(\mathbf{r}) - \nabla\phi_2(\mathbf{r})\cdot\nabla\phi_1(\mathbf{r})\right) d^3\mathbf{r} = 0 \tag{5.60}$$

by using per partes integration. The fact that $\hat{H}$ is symmetric can be denoted as follows: $\hat{H}^T = \hat{H}$ with $\hat{H}^T$ the transpose of $\hat{H}$. The Hamiltonian $\hat{H}$ is also called *self-adjoint* because it is both hermitian and symmetric. In symbols one can write $\hat{H}^+ = \hat{H}$ with $\hat{H}^+ = (\hat{H}^*)^T$ the adjoint of $\hat{H}$.

5.6.2 Orthogonality of Eigenfunctions

Let us consider two eigenfunctions $\phi_{\mathbf{n}}(\mathbf{r})$ and $\phi_{\mathbf{n}'}(\mathbf{r})$ which correspond to two different eigenvalues $E_{\mathbf{n}} \neq E_{\mathbf{n}'}$ of the quantum Hamiltonian $\hat{H}$. It is not difficult to prove that $\phi_{\mathbf{n}}(\mathbf{r})$ and $\phi_{\mathbf{n}'}(\mathbf{r})$ are *orthogonal*, i.e.

$$\int_{\mathbb{R}^3} \phi_{\mathbf{n}'}^*(\mathbf{r})\phi_{\mathbf{n}}(\mathbf{r})\, d^3\mathbf{r} = 0 \tag{5.61}$$

if $\mathbf{n} \neq \mathbf{n}'$. In fact, starting from

$$\hat{H}\phi_{\mathbf{n}}(\mathbf{r}) = E_{\mathbf{n}}\,\phi_{\mathbf{n}}(\mathbf{r})\,, \tag{5.62}$$

$$\hat{H}\phi_{\mathbf{n}'}^*(\mathbf{r}) = E_{\mathbf{n}'}\,\phi_{\mathbf{n}'}^*(\mathbf{r})\,, \tag{5.63}$$

we can write

$$\phi_{n'}^{*}(\mathbf{r})\hat{H}\phi_{n}(\mathbf{r}) = E_{n}\,\phi_{n'}^{*}(\mathbf{r})\,\phi_{n}(\mathbf{r})\,, \tag{5.64}$$

$$\phi_{n}(\mathbf{r})\hat{H}\phi_{n'}^{*}(\mathbf{r}) = E_{n'}\,\phi_{n}(\mathbf{r})\,\phi_{n'}^{*}(\mathbf{r})\,. \tag{5.65}$$

Subtracting the last two equations and integrating over space we get

$$\int_{\mathbb{R}^3}\phi_{n'}^{*}(\mathbf{r})\hat{H}\phi_{n}(\mathbf{r})\,d^3\mathbf{r} - \int_{\mathbb{R}^3}\phi_{n}(\mathbf{r})\hat{H}\phi_{n'}^{*}(\mathbf{r})\,d^3\mathbf{r} = (E_{n'} - E_{n})\int_{\mathbb{R}^3}\phi_{n'}^{*}(\mathbf{r})\phi_{n}(\mathbf{r})\,d^3\mathbf{r}\,. \tag{5.66}$$

The left side of this equality is zero because $\hat{H}$ is symmetric, and consequently

$$0 = (E_{n'} - E_{n})\int_{\mathbb{R}^3}\phi_{n'}^{*}(\mathbf{r})\phi_{n}(\mathbf{r})\,d^3\mathbf{r}\,. \tag{5.67}$$

Finally, because we have assumed that $E_{n'} \neq E_{n}$, we get Eq. (5.61).

Usually the eigenfunctions are normalized to one and we can then say that they are *orthonormal*, i.e.

$$\int_{\mathbb{R}^3}\phi_{n'}^{*}(\mathbf{r})\phi_{n}(\mathbf{r})\,d^3\mathbf{r} = \delta_{n',n} \tag{5.68}$$

with $\delta_{n',n}$ the Kronecher delta, such that $\delta_{n',n} = 1$ if $\mathbf{n} = \mathbf{n}'$ and $\delta_{n',n} = 0$ if $\mathbf{n} \neq \mathbf{n}'$.

Further Reading

Relevant historical papers about the De Broglie wavelength, the Schrödinger equation and its probabilistic interpretation are:

L. De Broglie, Annales de Physique **10**, 22 (1925).

E. Schrödinger, Physical Review **28**, 1049 (1926).

M. Born, Zeitschrift für Physik **37**, 863 (1926).

M. Born, Zeitschrift für Physik, **38**, 803 (1926).

E. Madelung, Naturwissenschaften **14**, 1004 (1926).

The cited works about the experimental verification of the wave-particle duality of the electron are:

C.J. Davisson and L.H. Germer, Proceedings National Academy of Sciences **14**, 317 (1928).

P.G. Merli, G.F. Missiroli, and G. Pozzi, American Journal of Physics **44**, 306 (1976).

A. Tonomura, J. Endo, T. Matsuda, T. Kawasaki, and H Ezawa, Americal Journal of Physics **57** 117 (1989).

A very nice pedagogical book on quantum mechanics and the Schrödinger wavefunction is:

R.W. Robinett, Quantum Mechanics: Classical Results, Modern Systems, and Visualized Examples (Oxford Univ. Press, 2006).

Chapter 6
Axiomatization of Quantum Mechanics

In this chapter we discuss the matrix mechanics of Born, Jordan, and Heisenberg, and then we analyze the basic axioms of quantum mechanics, which were formulated in the seminal books of Dirac and von Neumann. Finally, we consider the quantum perturbation theory for both time-independent and time-dependent cases.

6.1 Matrix Mechanics and Commutation Rules

In 1925, that is a year before the discovery of the Schrödinger equation, Max Born, Pasqual Jordan and Werner Heisenberg introduced the matrix mechanics. This was the first complete (in the non-relativistic limit) and coherent version of quantum mechanics and extended Bohr's atomic model of Bohr, justifying from the theoretical point of view the existence of the quantum jumps. This result was achieved by describing the physical observables observables and their time evolution through the use of matrices. The idea of matrix mechanics is to eliminate the concept of classical trajectory of elementary particle. Heisenberg wrote: "All my efforts are directed towards the unraveling and replacement of the concept of the orbital trajectory that one cannot observe". It is possible to show that the mechanics of the matrices of Heisenberg-Born-Jordan matrix mechanics is equivalent to the *wave mechanics* of Erwin Schrödinger, who followed it by about six months. The wave mechanics is more intuitive and mathematically simpler than the matrix mechanics.

According to the matrix mechanics the position **r** and the linear momentum **p** of an elementary particle are not vectors composed of numbers but instead vectors composed of infinite dimensional matrices (operators) which satisfy strange commutation rules, i.e.

$$\hat{\mathbf{r}} = (\hat{x}_1, \hat{x}_2, \hat{x}_3)\,, \qquad \hat{\mathbf{p}} = (\hat{p}_1, \hat{p}_2, \hat{p}_3)\,, \qquad (6.1)$$

© The Author(s), under exclusive license to Springer Nature Switzerland AG 2022
L. Salasnich, *Modern Physics*, UNITEXT for Physics,
https://doi.org/10.1007/978-3-030-93743-0_6

such that

$$[\hat{x}_j, \hat{p}_k] = i\hbar\,\delta_{jk}\,, \qquad (6.2)$$

where the hat symbol is introduced to denote operators, $[\hat{A}, \hat{B}] = \hat{A}\hat{B} - \hat{B}\hat{A}$ is the commutator of generic operators $\hat{A}$ and $\hat{B}$, and δ_{jk} is the Kronecker delta ($\delta_{jk} = 1$ if $j = k$ and $\delta_{jk} = 0$ if $j \neq k$). As usual, $i = \sqrt{-1}$ is the imaginary unit. In full generality, these infinite dimensional matrices (operators) act on infinite dimensional vectors (wavefunctions) which characterize the possible "quantum states" of the system.

By using their theory Born, Jordan and Heisenberg were able to obtain the energy spectrum of the hydrogen atom and also to calculate the transition probabilities between two energy levels. Soon after, Schrödinger realized that the matrix mechanics is equivalent to his wave-like formulation in the *coordinate representation* introducing the quantization rules

$$\hat{\mathbf{r}} = \mathbf{r}\,, \qquad \hat{\mathbf{p}} = -i\hbar\nabla\,. \qquad (6.3)$$

For instance, given a generic function $f(x_j)$ one finds immediately the commutation rule

$$\begin{aligned}
\left(\hat{x}_j\,\hat{p}_j - \hat{p}_j\,\hat{x}_j\right) f(x_j) &= -i\hbar x_j\,\frac{\partial}{\partial x_j}f(x_j) + i\hbar\,\frac{\partial}{\partial x_j}\left(x_j\,f(x_j)\right) \\
&= -i\hbar x_j\,\frac{\partial}{\partial x_j}f(x_j) + i\hbar\,f(x_j) + i\hbar x_j\,\frac{\partial}{\partial x_j}f(x_j) \\
&= i\hbar\,f(x_j)\,. \qquad (6.4)
\end{aligned}$$

Moreover, starting from classical Hamiltonian

$$H = \frac{p^2}{2m} + U(\mathbf{r})\,, \qquad (6.5)$$

the quantization rules give immediately the quantum Hamiltonian operator in the coordinate representation

$$\hat{H} = -\frac{\hbar^2}{2m}\nabla^2 + U(\mathbf{r}) \qquad (6.6)$$

from which one can write the time-dependent Schrödinger equation as

$$i\hbar\frac{\partial}{\partial t}\psi(\mathbf{r}, t) = \hat{H}\psi(\mathbf{r}, t)\,. \qquad (6.7)$$

6.1.1 Momentum Representation

Actually, one can also work in the so-called *momentum representation*, introducing the alternative quantization rules

$$\hat{\mathbf{r}} = i\hbar\nabla_{\mathbf{p}}, \qquad \hat{\mathbf{p}} = \mathbf{p}, \tag{6.8}$$

with $\nabla_{\mathbf{p}} = (\frac{\partial}{\partial p_1}, \frac{\partial}{\partial p_2}, \frac{\partial}{\partial p_3})$. For instance, given a generic function $f(p_j)$ one finds immediately the commutation rule

$$\begin{aligned}
\left(\hat{x}_j\,\hat{p}_j - \hat{p}_j\,\hat{x}_j\right) f(p_j) &= i\hbar\frac{\partial}{\partial p_j}\left(p_j\,f(p_j)\right) - i\hbar\,p_j\frac{\partial}{\partial p_j}f(p_j) \\
&= i\hbar\,f(p_j) + i\hbar\frac{\partial}{\partial p_j}\,f(p_j) - i\hbar\,p_j\frac{\partial}{\partial p_j}f(p_j) \\
&= i\hbar\,f(p_j)\,.
\end{aligned} \tag{6.9}$$

Moreover, starting from the classical Hamiltonian

$$H = \frac{p^2}{2m} + U(\mathbf{r})\,, \tag{6.10}$$

the quantization rules in the momentum representation give immediately the quantum Hamiltonian operator in the momentum representation

$$\hat{H} = \frac{p^2}{2m} + U(i\hbar\nabla_{\mathbf{p}}) \tag{6.11}$$

from which one can write the time-dependent Schrödinger equation

$$i\hbar\frac{\partial}{\partial t}\psi(\mathbf{p}, t) = \hat{H}\psi(\mathbf{p}, t)\,, \tag{6.12}$$

for the wavefunction $\psi(\mathbf{p}, t) = \psi(p_1, p_2, p_3, t)$.

6.2 Time Evolution Operator

A crucial operator in quantum mechanics is the *time evolution operator*, defined as

$$\hat{U}(t) = \exp\left(-i\hat{H}t/\hbar\right) = \sum_{n=0}^{+\infty}\frac{1}{n!}\left(-\frac{i}{\hbar}\hat{H}t\right)^n\,, \tag{6.13}$$

with $\hat{H}$ given by Eq. (6.6). Quite remarkably, the equation

$$\psi(\mathbf{r}, t) = \hat{U}(t)\,\psi(\mathbf{r}, 0) \tag{6.14}$$

is equivalent to Eq. (6.7). In fact, performing the time derivative in Eq. (6.14) we obtain

$$\frac{\partial}{\partial t}\psi(\mathbf{r}, t) = \frac{\partial}{\partial t}\left(\hat{U}(t)\,\psi(\mathbf{r}, 0)\right) = \left(\frac{\partial}{\partial t}\exp\left(-i\hat{H}t/\hbar\right)\right)\psi(\mathbf{r}, 0)$$

$$= -\frac{i}{\hbar}\,\hat{H}\,\exp\left(-i\hat{H}t/\hbar\right)\psi(\mathbf{r}, 0) = -\frac{i}{\hbar}\,\hat{H}\,\psi(\mathbf{r}, t)\,. \tag{6.15}$$

Both Eqs. (6.7) and (6.14) describe the time evolution of the wavefunction $\psi(\mathbf{r}, t)$.

6.3   Axioms of Quantum Mechanics

The axiomatic formulation of quantum mechanics was set up by Paul Maurice Dirac in 1930 and John von Neumann in 1932. The basic axioms are the following:

Axiom 1. The "state" of a quantum system is described by a vector $|\psi\rangle$ belonging to a complex Hilbert space $\mathcal{H}$. This state is usually called "ket ψ".

A complex Hilbert space $\mathcal{H}$ is a vector space, which can be finite dimensional or infinite dimensional, equipped with the complex scalar product (also called inner product) $\langle\psi|\psi'\rangle$ between any pair of states $|\psi\rangle, |\psi'\rangle \in \mathcal{H}$. The norm, or modulus, of a generic vector $|\psi\rangle \in \mathcal{H}$ is defined as

$$||\psi|| = |\langle\psi|\psi\rangle|\,,$$

and usually $|\psi\rangle$ is normalized to one, i.e. $||\psi|| = 1$. The symbol $\langle\psi|$ which appears in the definition of the norm is called "bra ψ" and it can be interpeted as the fuction $\langle\psi| :$ $\mathcal{H} \to \mathbb{C}$. For any $|\psi'\rangle \in \mathcal{H}$ this function gives a complex number $\langle\psi|\psi'\rangle$ obtained as scalar product of $|\psi\rangle$ and $|\psi'\rangle$. In a complex Hilbert space $\mathcal{H}$ it exists a set of basis vectors $|\phi_\alpha\rangle$ which are orthonormal, i.e. $\langle\phi_\alpha|\phi_\alpha\rangle = \delta_{\alpha,\beta}$, and such that

$$|\psi\rangle = \sum_\alpha c_\alpha|\phi_\alpha\rangle$$

for any $|\psi\rangle \in \mathcal{H}$, where the coefficients c_α belong to $\mathbb{C}$.

Axiom 2. Any observable (measurable quantity) of a quantum system is described by a self-adjoint linear operator $\hat{F}$ acting on the Hilbert space of state vectors.

For any classical observable F it exists a corresponding quantum observable $\hat{F} : \mathcal{H} \to \mathcal{H}$, which can be a finite-dimensional or infinite-dimensional matrix. Self-adjoint means that $\hat{F}^+ = \hat{F}$, where $\hat{F}^+$ is the adjoint (also called Hermitian conjugate or transpose conjugate) of the operator $\hat{F}$, i.e. $\hat{F}^+ = (\hat{F}^*)^t$ where $*$ means the complex conjugate and ^{t} means the transpose of the matrix.

Axiom 3. The possible measurable values of an observable $\hat{F}$ are its eigenvalues f, such that

$$\hat{F}|f\rangle = f|f\rangle$$

with $|f\rangle$ the corresponding eigenstate.

The observable $\hat{F}$ admits the spectral resolution

$$\hat{F} = \sum_f f\,|f\rangle\langle f|,$$

where $\{|f\rangle\}$ is the set of orthonormal eigenstates of $\hat{F}$, and the mathematical object $\langle f|$, called "bra of f", is a linear map that maps into the complex number. This spectral resolution is quite useful in applications, as well as the spectral resolution of the identity

$$\hat{I} = \sum_f |f\rangle\langle f|.$$

Axiom 4. The probability $\mathcal{P}$ of finding the state $|\psi\rangle$ in the state $|f\rangle$ is given by

$$\mathcal{P} = |\langle f|\psi\rangle|^2,$$

where the complex probability amplitude $\langle f|\psi\rangle$ denotes the scalar product of the two vectors. This probability $\mathcal{P}$ is also the probability of measuring the value f of the observable $\hat{F}$ when the system is in the quantum state $|\psi\rangle$.

In Axiom 4 both $|\psi\rangle$ and $|f\rangle$ must be normalized to one. Clearly, from Axiom 1 and Axiom 4 we have $|\psi\rangle = \sum_f c_f|f\rangle$, where c_f is the complex amplitude of the probability $\mathcal{P} = |\langle f|\psi\rangle|^2$ that is equal to $|c_f|^2$. Notice that the *wavefunction collapse* occurs when, after an observation (measurement), the quantum state $|\psi\rangle$ collapses into one of its basis states. Often it is useful to introduce the expectation value (mean value or average value) of an observable $\hat{F}$ with respect to a state $|\psi\rangle$ as $\langle\psi|\hat{F}|\psi\rangle$. Moreover, from the Axiom 4 it follows that the wavefunction $\psi(\mathbf{r})$ can be interpreted as

$$\psi(\mathbf{r}) = \langle\mathbf{r}|\psi\rangle,$$

that is the probability amplitude of finding the state $|\psi\rangle$ at the position state $|\mathbf{r}\rangle$, and similarly

$$\psi^*(\mathbf{r}) = \langle\psi|\mathbf{r}\rangle.$$

One can also consider $\psi(\mathbf{r})$ as the vector of a specific Hilbert space $\mathcal{H}$ that is usually denoted as $L^2(\mathbb{R}^3, \mathbb{C})$. This is the space of all the functions $\psi(\mathbf{r}) : \mathbb{R}^3 \to \mathbb{C}$ whose square modulus $|\psi(\mathbf{r})|^2$ is Lebesgue integrable. Within this interpretation, the scalar product of $|\psi\rangle$ and $|\phi\rangle$ can be written as

$$\langle\psi|\phi\rangle = \int_{\mathbb{R}^3} \psi^*(\mathbf{r})\phi(\mathbf{r})\,d^3\mathbf{r}\,,$$

and the norm $||\psi||$ of $|\psi\rangle$ is such that

$$||\psi||^2 = \int_{\mathbb{R}^3} |\psi(\mathbf{r})|^2\,d^3\mathbf{r}\,.$$

Axiom 5. The time evolution of states and observables of a quantum system with Hamiltonian $\hat{H}$ is determined by the unitary operator

$$\hat{U}(t) = \exp{(-i\hat{H}t/\hbar)}\,,$$

such that $|\psi(t)\rangle = \hat{U}(t)|\psi\rangle$ is the time-evolved state $|\psi\rangle$ at time t and $\hat{F}(t) = \hat{U}^{-1}(t)\hat{F}\hat{U}(t)$ is the time-evolved observable $\hat{F}$ at time t.

From Axiom 5, simply using the derivative wih respect to time t, one finds immediately the Schrödinger equation

$$i\hbar\frac{\partial}{\partial t}|\psi(t)\rangle = \hat{H}|\psi(t)\rangle$$

for the state $|\psi(t)\rangle$, and the Heisenberg equation

$$i\hbar\frac{\partial}{\partial t}\hat{F}(t) = [\hat{F}(t), \hat{H}]$$

for the observable $\hat{F}(t)$. In the so-called *Heisenberg picture*, that is the one of matrix mechanics, the states are time independent while the observables are time dependent. Instead, in the so-called *Schrödinger picture*, that is the one of the wave mechanics, the states are time dependent while the observables are, usually, time independent. Indeed, for the sake of simplicity, in the Heisenberg equation of the observable $\hat{F}(t)$ we are assuming that the observable is time-independent in the Schrödinger picture, i.e. $\hat{F}_S(t) = \hat{F}_S(0)$ where the subscript S means "Schrödinger picture". It is then quite intuitive to recognize that one can move from the Heisenberg picture to the Schrödinger picture by applying the time-evolution unitary operator $\hat{U}(t)$. Finally, we stress that this observable $\hat{F}$ is a *constant of motion* if it commutes with the Hamiltonian operator $\hat{H}$.

6.4   Heisenberg Uncertainty Principle

The position-momentum uncertainty principle, introduced by Werner Heisenberg in 1927, is a theorem of quantum mechanics which follows from the commutation rule (6.2) of the observables position $\hat{\mathbf{r}} = (\hat{x}_1, \hat{x}_2, \hat{x}_3)$ and linear momentum $\hat{\mathbf{p}} = (\hat{p}_1, \hat{p}_2, \hat{p}_3)$, as shown by Earle Hesse Kennard in 1927 and by Hermann Weyl in 1928. The principle says that the standard deviations Δx_j and Δp_j always satisfy the inequality

$$\Delta x_j \, \Delta p_j \geq \frac{\hbar}{2}, \tag{6.16}$$

where

$$(\Delta x_j)^2 = \langle \psi | \hat{x}_j^2 | \psi \rangle - \langle \psi | \hat{x}_j | \psi \rangle^2 \tag{6.17}$$

and

$$(\Delta p_j)^2 = \langle \psi | \hat{p}_j^2 | \psi \rangle - \langle \psi | \hat{p}_j | \psi \rangle^2, \tag{6.18}$$

with $|\psi\rangle$ a generic quantum state.

Actually, the position-momentum inequality (6.16) is nothing else than the position-wavevector theorem

$$\Delta x_j(t) \, \Delta k_j(t) \geq \frac{1}{2}, \tag{6.19}$$

of the space-wavevector Fourier transform

$$\tilde{\psi}(\mathbf{k}, t) = \int \psi(\mathbf{r}, t) \, e^{-i\mathbf{k}\cdot\mathbf{r}} \, d^3\mathbf{r} \tag{6.20}$$

of the wavefunction $\psi(\mathbf{r}, t)$, observing that $\hat{\mathbf{p}} = \hbar\hat{\mathbf{k}}$ with $\hat{\mathbf{k}} = -i\nabla$ and that

$$(\Delta p_j(t))^2 = \hbar^2 \, \Delta k_j(t)^2, \tag{6.21}$$

where

$$(\Delta k_j(t))^2 = \langle \psi | \hat{k}_j^2 | \psi \rangle - \langle \psi | \hat{k}_j | \psi \rangle^2$$

$$= -\int \psi^*(\mathbf{r}, t) \frac{\partial^2}{\partial x_j^2} \psi(\mathbf{r}, t) \, d^3\mathbf{r} - \left(i \int \psi^*(\mathbf{r}, t) \frac{\partial}{\partial x_j} \psi(\mathbf{r}, t) \, d^3\mathbf{r} \right)^2$$

$$= \int k_j^2 \, |\tilde{\psi}(\mathbf{k}, t)|^2 \, d^3\mathbf{k} - \left(\int k_j \, |\tilde{\psi}(\mathbf{k}, t)|^2 \, d^3\mathbf{k} \right)^2. \tag{6.22}$$

For the sake of completeness, we observe that, given the time-frequency Fourier transform

$$\tilde{\psi}(\mathbf{r}, \omega) = \int_{-\infty}^{+\infty} \psi(\mathbf{r}, t)\, e^{i\omega t}\, dt \tag{6.23}$$

of the wavefunction $\psi(\mathbf{r}, t)$ in the frequency domain ω, the frequency-time theorem holds

$$\Delta\omega\, \Delta t \geq \frac{1}{2}, \tag{6.24}$$

where

$$(\Delta\omega)^2 = \int d^3\mathbf{r} \int_{-\infty}^{\infty} d\omega\, \omega^2\, |\tilde{\psi}(\mathbf{r}, \omega)|^2 - \left(\int d^3\mathbf{r} \int_{-\infty}^{\infty} d\omega\, \omega\, |\tilde{\psi}(\mathbf{r}, \omega)|^2\, d^3\mathbf{r}\, d\omega \right)^2 \tag{6.25}$$

and

$$(\Delta t)^2 = \int d^3\mathbf{r} \int_{-\infty}^{\infty} dt\, t^2\, |\psi(\mathbf{r}, t)|^2 - \left(\int d^3\mathbf{r} \int_{-\infty}^{\infty} dt\, t\, |\psi(\mathbf{r}, t)|^2 \right)^2. \tag{6.26}$$

Setting $E = \hbar\omega$ we immediately find that energy-time uncertainty relation

$$\Delta E\, \Delta t \geq \frac{\hbar}{2}. \tag{6.27}$$

In the simultaneous measurements of the energy of a quantum state and its lifetime, Eq. (6.27) connects the uncertainty ΔE in the energy measurement to the uncertainty Δt in the lifetime measurement. Indeed, as seen in Chap. 4, ΔE is strictly related to the natural linewidth of the quantum state. Within the Schrödinger picture, where the time t is identified as the parameter entering in the Schrödinger equation, the energy-time uncertainty is meaningful. However, within the Heisenberg picture, the energy-time uncertainty relation is a bit controversial. The self-adjoint operator associated to the energy E is the quantum Hamiltonian $\hat{H}$, while the existence of a time operator $\hat{t}$ gives rise to technical problems.

6.4.1 Uncertainty Principle for Non-commuting Operators

It is important to observe that there is a more formal version of the Heisenberg uncertainty principle. Let us consider two operators $\hat{A}$ and $\hat{B}$. Their mean-square uncertainties are given by

$$(\Delta A)^2 = \langle \hat{A}^2 \rangle - \langle \hat{A} \rangle^2, \tag{6.28}$$

$$(\Delta B)^2 = \langle \hat{B}^2 \rangle - \langle \hat{B} \rangle^2, \tag{6.29}$$

where $\langle \hat{A}^2 \rangle = \langle \psi | \hat{A}^2 | \psi \rangle$ and $\langle \hat{A} \rangle^2 = \langle \psi | \hat{A} | \psi \rangle^2$ with $|\psi\rangle$ a chosen quantum state. It is then possible to prove (Howard Percy Robertson, 1929) that

$$\Delta A \, \Delta B \geq \frac{1}{2} |\langle [\hat{A}, \hat{B}] \rangle| \, , \tag{6.30}$$

where $[\hat{A}, \hat{B}] = \hat{A}\hat{B} - \hat{B}\hat{A}$ is the commutator. If $[\hat{A}, \hat{B}] \neq 0$, Eq. (6.30) is the uncertainty principle for non-commuting operators. If $[\hat{A}, \hat{B}] = 0$ the two operators are commuting and Eq. (6.30) implies no uncertainty, i.e. $\Delta A \, \Delta B = 0$. Notice that, if commutator $[\hat{A}, \hat{B}]$ is a constant, the expectation value in Eq. (6.30) can be removed and we obtain the simplfied expression

$$\Delta A \, \Delta B \geq \frac{1}{2} |[\hat{A}, \hat{B}]| \, . \tag{6.31}$$

Clearly, with two operators $\hat{x}_j$ and $\hat{p}_j$, such that $[\hat{x}_j, \hat{p}_j] = i\hbar$, from Eq. (6.31) we find exactly Eq. (6.16).

Two quantum observabiles $\hat{A}$ and $\hat{B}$ are called *complementary* if they are not commuting, i.e. $[\hat{A}, \hat{B}] \neq 0$. The observables are instead called *compatible* if they commute, i.e. $[\hat{A}, \hat{B}] = 0$. Quite remarkably, complementary observables cannot have common eigenstates, while compatible observables may have common eigenstates.

6.5 Time-Independent Perturbation Theory

Let us consider a quantum system described by the Hamiltonian

$$\hat{H} = \hat{H}_0 + \hat{H}_I \, , \tag{6.32}$$

where $\hat{H}_0$ is an Hamiltonian whose eigenvalues and eigenstates are exactly known, while $\hat{H}_I$ is simply the difference between $\hat{H}$ and $\hat{H}_0$. A generic eigenstate $|\phi_n^{(0)}\rangle$ of the Hamiltonian $\hat{H}_0$ satisfies the stationary Schrödinger equation

$$\hat{H}_0 |\phi_n^{(0)}\rangle = E_n^{(0)} |\phi_n^{(0)}\rangle \, . \tag{6.33}$$

Our aim is to solve the eigenvalue problem

$$\hat{H} |\phi_n\rangle = E_n |\phi_n\rangle \, , \tag{6.34}$$

in a perturbative way. For this reason it is useful to introduce a parameter $\lambda \in [0, 1]$ used to keep track of the order of the correction. In particular, instead of Eq. (6.32), we consider

$$\hat{H} = \hat{H}_0 + \lambda \hat{H}_I \, . \tag{6.35}$$

H_0 is usually called unperturbed Hamiltonian while $\lambda\,H_I$ is the perturbing Hamiltonian. Clearly $\lambda = 0$ means no perturbation and $\lambda = 1$ means full perturbation. At the end of the calculations, at a fixed order of the perturbation theory, one sets $\lambda = 1$. Obviously one must also verify that the obtained corrections are small with respect to the unperturbed ones. The perturbative procedure, that is called Rayleigh-Schrödinger perturbation theory, is then based on the following λ expansions for eigenvalues and eigenstates:

$$E_{\mathbf{n}} = \sum_{j=0}^{+\infty} \lambda^j\, E_{\mathbf{n}}^{(j)} = E_{\mathbf{n}}^{(0)} + \lambda\, E_{\mathbf{n}}^{(1)} + \lambda^2\, E_{\mathbf{n}}^{(2)} + \dots \,, \tag{6.36}$$

$$|\phi_{\mathbf{n}}\rangle = \sum_{j=0}^{+\infty} \lambda^j |\phi_{\mathbf{n}}^{(j)}\rangle = |\phi_{\mathbf{n}}^{(0)}\rangle + \lambda\,|\phi_{\mathbf{n}}^{(1)}\rangle + \lambda^2\,|\phi_{\mathbf{n}}^{(2)}\rangle + \dots \,. \tag{6.37}$$

By using Eqs. (6.35) and (6.37) we immediately obtain

$$E_{\mathbf{n}} = \langle\phi_{\mathbf{n}}|\hat{H}|\phi_{\mathbf{n}}\rangle = \langle\phi_{\mathbf{n}}|\left(\hat{H}_0 + \lambda\hat{H}_I\right)|\phi_{\mathbf{n}}\rangle = E_{\mathbf{n}}^{(0)} + \lambda\,\langle\phi_{\mathbf{n}}^{(0)}|\hat{H}|\phi_{\mathbf{n}}^{(0)}\rangle + \lambda^2\langle\phi_{\mathbf{n}}^{(1)}|\hat{H}_0|\phi_{\mathbf{n}}^{(1)}\rangle + \dots \,. \tag{6.38}$$

Consequently, setting $\lambda = 1$, at the first order of the perturbation theory we simply get

$$E_{\mathbf{n}} = E_{\mathbf{n}}^{(0)} + \langle\phi_{\mathbf{n}}^{(0)}|\hat{H}_I|\phi_{\mathbf{n}}^{(0)}\rangle \,. \tag{6.39}$$

This is a very nice result: the eigenvalues of the total Hamiltonian are obtained knowing the eigenvalues of the unperturbed Hamiltonian and the average (expectation value) of the perturbing Hamiltonian with respect to the unperturbed eigenstates. Thus, the perturbation causes the average energy of this state to change by $E_{\mathbf{n}}^{(1)} = \langle\phi_{\mathbf{n}}^{(0)}|\hat{H}_I|\phi_{\mathbf{n}}^{(0)}\rangle$. This is the so-called first-order result of non-degenerate perturbation theory for the energy levels. Equation (6.39) can be extendend to the degenerate case, where various unperturbed eigenstates $|\psi_{\mathbf{n},\alpha}^{(0)}\rangle$ correspond to the same unperturbed eigenvalue $E_{\mathbf{n}}^{(0)}$. Here the label α takes into account the degeneration, i.e. $\alpha = 1, 2, \dots, M$ for a M times degenerate eigenvalue. In this case, the first-order degenerate perturbation theory gives

$$E_{\mathbf{n},\alpha} = E_{\mathbf{n}}^{(0)} + E_{\mathbf{n},\alpha}^{(1)} \,, \tag{6.40}$$

where the various $E_{\mathbf{n},\alpha}^{(1)}$ are the eigenvalues of the $M \times M$ perturbing matrix whose elements are $\langle\phi_{\mathbf{n},\alpha}^{(0)}|\hat{H}_I|\phi_{\mathbf{n},\beta}^{(0)}\rangle$.

The derivation of the first-order correction for the eigenstates is quite complicated also in the non-degenerate case. Here we report the final formula:

$$|\phi_{\mathbf{n}}^{(1)}\rangle = \sum_{n'\neq n} \frac{\langle\phi_{\mathbf{n}'}^{(0)}|\hat{H}_I|\phi_{\mathbf{n}}^{(0)}\rangle}{E_{\mathbf{n}}^{(0)} - E_{\mathbf{n}'}^{(0)}}\, |\phi_{\mathbf{n}'}^{(0)}\rangle \,. \tag{6.41}$$

From this expression one can then deduce the second-order correction for the eigenvalues, which reads

$$E_{\mathbf{n}}^{(2)} = \sum_{n' \neq n} \frac{|\langle \phi_{\mathbf{n}'}^{(0)} | \hat{H}_I | \phi_{\mathbf{n}}^{(0)} \rangle|^2}{E_{\mathbf{n}}^{(0)} - E_{\mathbf{n}'}^{(0)}} . \tag{6.42}$$

Note that in the last two formulas the summation is done with respect to $\mathbf{n}'$ with the constraint that $\mathbf{n}'$ must be different from $\mathbf{n}$.

6.6 Time-Dependent Perturbation Theory and Fermi Golden Rule

Let us consider again the Hamiltonian (6.32) equipped by Eq. (6.33). If the perturbation is zero, i.e. if $\hat{H}_I = 0$, then the time evolution of $|\phi_{\mathbf{n}}\rangle$ is simple:

$$|\phi_{\mathbf{n}}(t)\rangle = e^{-i\varepsilon_n t/\hbar} |\phi_{\mathbf{n}}(0)\rangle . \tag{6.43}$$

Clearly, in this case, the probability of finding the eigenstate $|\phi_{\mathbf{n}}\rangle$ of the unperturbed Hamiltonian $\hat{H}_0$ in another eigenstate $|\psi_l\rangle$ of the unperturbed Hamiltonian $\hat{H}_0$ is zero.

If instead the perturbation is not zero, i.e. if $\hat{H}_I \neq 0$, then the time evolution of $|\phi_{\mathbf{n}}\rangle$ is, in general, quite complicated because, usually, $|\phi_{\mathbf{n}}\rangle$ is not an eigenstate of the total Hamiltonian $\hat{H}$. The Fermi golden rule is relevant in this case because it gives a way to calculate the probability of finding the eigenstate $|\phi_{\mathbf{n}}\rangle$ of the unperturbed Hamiltonian $\hat{H}_0$ into another eigenstate $|\phi_l\rangle$ of the unperturbed Hamiltonian $\hat{H}_0$.

A generic time-dependent state $|\psi(t)\rangle$ satisfies the time-dependent Schrödinger equation

$$i\hbar \frac{\partial}{\partial t} |\psi(t)\rangle = \left(\hat{H}_0 + \hat{H}_I \right) |\psi(t)\rangle . \tag{6.44}$$

This state $|\psi(t)\rangle$ can be expanded in the orthonormal basis of the time-independent eigenstates $|\phi_j(0)\rangle$ of the unperturbed Hamiltonian $\hat{H}_0$ as follows

$$|\psi(t)\rangle = \sum_j c_{\mathbf{j}}(t) \, e^{-i\varepsilon_j t/\hbar} |\phi_{\mathbf{j}}(0)\rangle , \tag{6.45}$$

where the complex coefficients $c_{\mathbf{j}}(t)$ are all equal to one only in the very special case of $\hat{H}_I = 0$.

For the sake of simplicity we approximate Eq. (6.45) adopting the two-mode approximation which involves only two eigenstates $|\phi_I(0)\rangle$ and $|\phi_F(0)\rangle$ of the unperturbed Hamiltonian H_0:

$$|\psi(t)\rangle = \sum_{j=I,F} c_{\mathbf{j}}(t) \, e^{-i\varepsilon_j t/\hbar} |\phi_{\mathbf{j}}(0)\rangle , \tag{6.46}$$

assuming that at $t = 0$ the state $|\psi(0)\rangle$ of the system is in the initial state $|\phi_I(0)\rangle$, namely $c_I(0) = 1$ and $c_F(0) = 0$. Here $|\phi_F(0)\rangle$ is our final state, and clearly $\langle\phi_I(0)|\phi_F(0)\rangle = 0$.

Inserting the expression (6.46) into Eq. (6.44) and the bra $\langle\phi_F(0)|$ on the left side of the resulting formula we obtain

$$i\hbar\,\dot{c}_F(t) = \langle\phi_F(0)|\hat{H}_I|\phi_I(0)\rangle\,e^{i\omega_{IF}t}\,, \tag{6.47}$$

where $\omega_{IF} = (E_I - E_F)/\hbar$. The solution of this equation is given by

$$c_F(t) = \frac{\langle F|\hat{H}_I|I(0)\rangle}{i\hbar}\int_0^t e^{i\omega_{IF}t'}\,dt' = \frac{\langle F|\hat{H}_I|I\rangle}{\hbar\omega_{IF}}\left(1 - e^{i\omega_{IF}t}\right)\,, \tag{6.48}$$

where we set $|I\rangle = |\phi_I(0)\rangle$ and $F = |\phi_F(0)\rangle$. It follows that

$$|c_F(t)|^2 = \frac{|\langle\phi_F(0)|\hat{H}_I|\phi_F(0)\rangle|^2}{\hbar^2\omega_{IF}^2}4\sin^2\left(\omega_{IF}t/2\right)\,. \tag{6.49}$$

We can now introduce the transition probability per unit time

$$\frac{|c_F(t)|^2}{t} = \frac{1}{\hbar^2}\,t\,\frac{\sin^2(\omega_{IF}t/2)}{(\omega_{IF}t/2)^2}\,. \tag{6.50}$$

Since the Dirac delta function $\delta(x)$ can be written as

$$\delta(x) = \frac{1}{2\pi}\lim_{t\to+\infty}t\,\frac{\sin^2(x\,t)}{(x\,t)^2}\,, \tag{6.51}$$

the asymptotic transition probability per unit time

$$W_{IF} = \lim_{t\to+\infty}\frac{|c_F(t)|^2}{t} \tag{6.52}$$

reads

$$W_{IF} = \frac{2\pi}{\hbar}|\langle F|H_I|I\rangle|^2\,\delta(E_I - E_F)\,, \tag{6.53}$$

which is the so-called Fermi golden rule. Actually, it was derived for the first time in 1926 by Paul Dirac and named "golden rule" few years later by Enrico Fermi. Note that the Dirac delta function is then removed by integrating over the possible final states $|F\rangle$ but, obviously, to perform the integration one needs an explicit expression for the final states.

6.7 Variational Principle

Many approaches to the determination of the ground-state of a quantum system are based on the so-called variational principle, which is actually a theorem.

Theorem 1 *For any normalized quantum state $|\phi\rangle$, i.e. such that $\langle\phi|\phi\rangle = 1$, which belongs to the Hilbert space on which acts the Hamiltonian $\hat{H}$, one finds*

$$\langle\phi|\hat{H}|\phi\rangle \geq E_{gs} , \tag{6.54}$$

where E_{gs} is the ground-state energy of the system and the equality holds only if $|\phi\rangle = |\phi_{gs}\rangle$ with $|\phi_{gs}\rangle$ ground-state of the system, i.e. such that $\hat{H}|\phi_{gs}\rangle = E_{gs}|\phi_{gs}\rangle$.

Proof The quantum Hamiltonian $\hat{H}$ satisfies the exact eigenvalue problem

$$\hat{H}|\phi_\alpha\rangle = E_\alpha|\phi_\alpha\rangle , \tag{6.55}$$

where E_α are the ordered eigenvalues, i.e. such that $E_0 < E_1 < E_2 < ...$ with $E_0 = E_{gs}$, and $|\phi_\alpha\rangle$ the corresponding orthonormalized eigenstates, i.e. such that $\langle\phi_\alpha|\phi_\beta\rangle = \delta_{\alpha,\beta}$ with $|\phi_0\rangle = |\phi_{gs}\rangle$. The generic quantum state $|\phi\rangle$ can be written as

$$|\phi\rangle = \sum_\alpha c_\alpha|\phi_\alpha\rangle , \tag{6.56}$$

where c_α are the complex coefficients of the expansion such that

$$\sum_\alpha |c_\alpha|^2 = 1 . \tag{6.57}$$

Then one finds

$$\langle\phi|\hat{H}|\phi\rangle = \sum_{\alpha,\beta} c_\alpha^* c_\beta \langle\phi_\alpha|\hat{H}|\phi_\beta\rangle = \sum_{\alpha,\beta} c_\alpha^* c_\beta E_\beta \langle\phi_\alpha|\phi_\beta\rangle = \sum_{\alpha,\beta} c_\alpha^* c_\beta E_\beta \delta_{\alpha,\beta}$$

$$= \sum_\alpha |c_\alpha|^2 E_\alpha \geq \sum_\alpha |c_\alpha|^2 E_0 = E_0 = E_{gs} . \tag{6.58}$$

Obviously, the equality holds only if $c_0 = 1$ and, consequently, all the other coefficients are zero.

As an application of the variational method we will calculate the approximate energy of the ground-state of a one-dimensional quantum particle under the action of the quartic potential

$$U(x) = A\, x^4 . \tag{6.59}$$

The stationary Schrödinger equation of the particle is given by

$$\left[-\frac{\hbar^2}{2m}\frac{\partial^2}{\partial x^2} + A\,x^4 \right]\phi(x) = \epsilon\,\phi(x)\,, \tag{6.60}$$

while the energy that we want to minimize reads

$$E = \langle\phi|\hat{H}|\phi\rangle = \int dx\,\phi^*(x)\left[-\frac{\hbar^2}{2m}\frac{\partial^2}{\partial x^2} + A\,x^4 \right]\phi(x) = \int dx\left[\frac{\hbar^2}{2m}\left|\frac{\partial\phi(x)}{\partial x}\right|^2 + A\,x^4|\phi(x)|^2 \right], \tag{6.61}$$

with the normalization condition

$$\int dx\,|\phi(x)|^2 = 1 \tag{6.62}$$

for the wavefunction. We assume that the variational wavefunction is a Gaussian function, i.e.

$$\phi(x) = \frac{e^{-x^2/(2\sigma^2)}}{\pi^{1/4}\sigma^{1/2}}\,, \tag{6.63}$$

where σ is the variational parameter. Inserting this variational wavefunction into the energy functional and integrating over x we get

$$E = \frac{\hbar^2}{4m\sigma^2} + \frac{3}{4}A\,\sigma^4\,. \tag{6.64}$$

Minimizing this energy with respect to the variational parameter σ, i.e. setting $\frac{dE}{d\sigma} = 0$, we find

$$\sigma = \left(\frac{\hbar^2}{6\,mA} \right)^{1/6}\,. \tag{6.65}$$

Substituting this value of σ in the energy E we finally obtain

$$E = \frac{9}{4}\left(\frac{\hbar^2}{6m} \right)^{2/3} A^{1/3}\,, \tag{6.66}$$

which is the approximate energy of the ground-state. This energy is surely larger or equal to the energy of the exact ground state of the system.

Further Reading

Relevant historical papers about the matrix mechanics are:
W. Heisenberg, Zeitschrift für Physik **33**, 879 (1925).
M. Born and P. Jordan, Zeitschrift für Physik **34**, 858 (1925).
M. Born, W. Heisenberg, and P. Jordan, Zeitschrift für Physik **35**, 557 (1926).

The two seminal books on the axiomatization of quantum mechanics are:
P.A.M. Dirac, The Principles of Quantum Mechanics. First published in 1930 (Oxford Univ. Press, 1982).
J. von Neumann, Mathematical Foundations of Quantum Mechanics. First published in 1932 (Princeton Univ. Press, 2018).
For a detailed discussion of the quantum perturbation theory and the variational principle:
J.J. Sakurai and J. Napolitano, Modern Quantum Mechanics (Cambridge Univ. Press, 2020).

Chapter 7
Solvable Problems in Quantum Mechanics

In this chapter we consider relevant applications of quantum mechanics in one spatial dimension. Among them, the quantum particle in a box, in a harmonic potential, and in a double-well potential. We also discuss the three-dimensional Schrödinger equation with a separable potential.

7.1 One-Dimensional Square-Well Potential

The stationary Schrödinger equation for a particle moving only along the x-axis is given by

$$\hat{H}\,\phi(x) = E\,\phi(x)\,, \qquad (7.1)$$

where

$$\hat{H} = -\frac{\hbar^2}{2m}\frac{d^2}{dx^2} + U(x) \qquad (7.2)$$

is the Hamiltonian operator of this one-dimensional (1D) problem.

We assume that the external potential $U(x)$ is an infinite square-well, i.e.

$$U(x) = \begin{cases} 0 & \text{if } 0 \le x \le L \\ +\infty & \text{elsewhere} \end{cases} \qquad (7.3)$$

Therefore the particle is free to move in the region of x between 0 and L, while it cannot enter in the regions external to this interval $[0, L]$ since there the potential barrier becomes infinitely repulsive. Basically the problem under consideration reduces to that described by the stationary Schrödinger equation

Fig. 7.1 Infinite square-well
potential $U(x)$ with $L = 1$

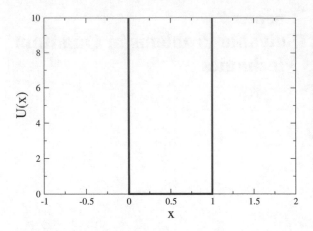

$$-\frac{\hbar^2}{2m}\frac{d^2}{dx^2}\phi(x) = E\,\phi(x) \tag{7.4}$$

with boundary conditions $\phi(0) = 0$ and $\phi(L) = 0$ (Fig. 7.1).

The equation can be rewritten as follows

$$\phi''(x) + k^2\phi(x) = 0, \tag{7.5}$$

where the constant k is given by

$$k = \sqrt{\frac{2mE}{\hbar^2}}. \tag{7.6}$$

This is a second order differential equation with constant coefficients. The general
solution is

$$\phi(x) = A\,e^{\kappa_1 x} + B\,e^{\kappa_2 x}, \tag{7.7}$$

where κ_1 and κ_2 are the roots of the algebraic equation

$$\kappa^2 + k^2 = 0, \tag{7.8}$$

namely

$$\kappa_1 = i\,k \qquad \kappa_2 = -i\,k. \tag{7.9}$$

The solution then becomes

$$\phi(x) = A\,e^{ikx} + B\,e^{-ikx}, \tag{7.10}$$

where the arbitrary constants A and B are determined via the boundary conditions
and by normalization. From the boundary conditions we have

$$0 = \phi(0) = A + B, \tag{7.11}$$

from which $A = -B$, but also

$$0 = \phi(L) = A e^{ikL} + B e^{-ikL} = A e^{ikL} - A e^{-ikL} = 2 i A \sin(kL). \tag{7.12}$$

It follows that

$$k L = \pi n \tag{7.13}$$

where n is an integer different from zero. We can then write

$$k = \frac{\pi}{L} n. \tag{7.14}$$

Recalling the definition of the constant k, which is precisely the wave number, we get

$$E_n = \frac{\hbar^2 \pi^2}{2mL^2} n^2, \tag{7.15}$$

which is the quantization formula for the energy levels of the problem. To recap, we obtained the eigenfunctions

$$\phi_n(x) = A \sin\left(\frac{\pi}{L} n x\right) \tag{7.16}$$

and the corresponding eigenvalues given by Eq. (7.15). It remains to determine the arbitrary constant A. Imposing the normalization condition

$$\int_0^L |\phi_n(x)|^2 \, dx = 1 \tag{7.17}$$

after some calculations one finds

$$A = \sqrt{\frac{2}{L}}. \tag{7.18}$$

7.2 One-Dimensional Harmonic Potential

In this section we assume that the external potential $U(x)$ is given by

$$U(x) = \frac{1}{2} m \omega^2 x^2 \tag{7.19}$$

This harmonic potential represents the potential energy of a particle of mass m subjected to the elastic force $F = -K_{el} x$, where K_{el} is the elastic constant and

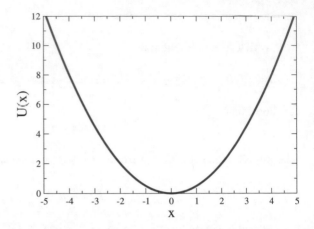

Fig. 7.2 Harmonic potential $U(x) = m\omega^2 x^2/2$ setting $m\omega^2 = 1$

$$\omega = \sqrt{\frac{K_{el}}{m}} \tag{7.20}$$

is the angular frequency of the periodic (harmonic) oscillation of the particle (Fig. 7.2).

The problem under consideration is described by the stationary Schrödinger equation

$$-\frac{\hbar^2}{2m}\frac{d^2}{dx^2}\phi(x) + \frac{1}{2}m\,\omega^2 x^2\,\phi(x) = E\,\phi(x) \tag{7.21}$$

with boundary conditions $\phi(-\infty) = 0$ e $\phi(+\infty) = 0$, which are needed to obtain a square-integrable smooth wavefunction. This is a second order differential equation but, unfortunately, it does not have constant coefficients.

To solve this equation we use a special approach that is the *factorization method*. The Hamiltonian operator of the problem at hand reads

$$\hat{H} = -\frac{\hbar^2}{2m}\frac{d^2}{dx^2} + \frac{1}{2}m\omega^2 x^2 = \frac{\hat{p}^2}{2m} + \frac{1}{2}m\omega^2 x^2 \,, \tag{7.22}$$

where $\hat{p} = -i\hbar d/dx$. By collecting $\hbar\omega$ this operator becomes

$$\hat{H} = \hbar\omega\left(-\frac{1}{2}\frac{\hbar}{m\omega}\frac{d^2}{dx^2} + \frac{1}{2}\frac{m\omega}{\hbar}x^2\right) = \hbar\omega\left(\frac{\hat{p}^2}{2m\hbar\omega} + \frac{1}{2}\frac{m\omega}{\hbar}x^2\right) \,. \tag{7.23}$$

Note that, using the characteristic length

$$\sigma = \sqrt{\frac{\hbar}{m\omega}} \tag{7.24}$$

appearing in the Hamiltonian, it can then be rewritten as

$$\hat{H} = \hbar\omega \left(-\frac{1}{2}\frac{d^2}{d\tilde{x}^2} + \frac{1}{2}\tilde{x}^2 \right) = \hbar\omega \left(\frac{1}{2}\hat{\tilde{p}}^2 + \frac{1}{2}\hat{\tilde{x}}^2 \right) \tag{7.25}$$

having introduced the dimensionless coordinate $\tilde{x} = x/\sigma$ and the dimensionless linear momentum operator $\hat{\tilde{p}} = -id/d\tilde{x}$. In 1927 Paul Dirac observed that the following formula holds

$$\hat{H} = \hbar\omega \left(\hat{a}^+ \hat{a} + \frac{1}{2} \right) , \tag{7.26}$$

where the differential operator $\hat{a}^+$ ed $\hat{a}$ are given by

$$\hat{a}^+ = \frac{1}{\sqrt{2}} \left(\tilde{x} - \frac{d}{d\tilde{x}} \right) = \left(\hat{\tilde{x}} + i\hat{\tilde{p}} \right) , \tag{7.27}$$

$$\hat{a} = \frac{1}{\sqrt{2}} \left(\hat{\tilde{x}} + \frac{d}{d\tilde{x}} \right) = \left(\hat{\tilde{x}} - i\hat{\tilde{p}} \right) . \tag{7.28}$$

Clearly, one can also write the inverse transformations

$$\hat{\tilde{x}} = \frac{1}{\sqrt{2}} \left(\hat{a} + \hat{a}^+ \right) , \tag{7.29}$$

$$\hat{\tilde{p}} = \frac{1}{i\sqrt{2}} \left(\hat{a} - \hat{a}^+ \right) . \tag{7.30}$$

Thus, in the case of the harmonic oscillator, the Hamiltonian operator $\hat{H}$ can be factorized in terms of the operators $\hat{a}^+$ and $\hat{a}$. The operators $\hat{a}^+$ and $\hat{a}$ satisfy the commutation rule

$$[\hat{a}, \hat{a}^+] = 1 , \tag{7.31}$$

i.e. explicitly

$$\hat{a}\,\hat{a}^+ - \hat{a}^+\,\hat{a} = 1 . \tag{7.32}$$

This commutation rule can obviously derived from the commutation rule $[\hat{x}, \hat{p}] = i\hbar$ or, equivalently, $[\hat{\tilde{x}}, \hat{\tilde{p}}] = i$. The operator $\hat{a}^+$ is called the creation operator, the operator $\hat{a}$ is called annihilation operator. These operators are also called *ladder operators*. Their product $\hat{a}^+\hat{a}$ is the *number operator*, denoted by the symbol $\hat{N}$, i.e.

$$\hat{N} = \hat{a}^+\,\hat{a} . \tag{7.33}$$

The reasons for these distinctive names are discussed below.

For the number operator it can be proved that the following eigenvalue equation holds

$$\hat{N}\,\phi_n(\tilde{x}) = n\,\phi_n(\tilde{x}) , \tag{7.34}$$

where the eigenvalue is just the natural number n and the corresponding eigenfunction $\phi_n(\tilde{x})$ is given by

$$\phi_n(\tilde{x}) = \frac{1}{\sqrt{n!}} \left(\hat{a}^+\right)^n \phi_0(\tilde{x}) , \tag{7.35}$$

where

$$\phi_0(\tilde{x}) = \frac{1}{\pi^{1/4}} e^{-\tilde{x}^2/2} . \tag{7.36}$$

Furthermore, the operators $\hat{a}$ and $\hat{a}^+$ are such that

$$\hat{a} \, \phi_n(\tilde{x}) = \sqrt{n} \, \phi_{n-1}(\tilde{x}) , \tag{7.37}$$
$$\hat{a}^+ \phi_n(\tilde{x}) = \sqrt{n+1} \, \phi_{n+1}(\tilde{x}) . \tag{7.38}$$

These formulas justify the names of destruction and creation operators for $\hat{a}$ and $\hat{a}^+$. Note that, adopting the Dirac notation, Eq. (7.34) can be written in a more compact way as

$$\hat{N} \, |n\rangle = n \, |n\rangle . \tag{7.39}$$

Similarly, Eqs. (7.38) and (7.38) can be written as

$$\hat{a} \, |n\rangle = \sqrt{n} \, |n-1\rangle , \tag{7.40}$$
$$\hat{a}^+ |n\rangle = \sqrt{n+1} \, |n+1\rangle . \tag{7.41}$$

In conclusion, since

$$\hat{H} = \hbar\omega \left(\hat{N} + \frac{1}{2}\right) , \tag{7.42}$$

it follows

$$\hat{H} \, \phi_n(x) = \hbar\omega \left(n + \frac{1}{2}\right) \phi_n(x) , \tag{7.43}$$

with $E_n = \hbar\omega(n + 1/2)$ a generic quantized energy level with quantum number n and with $\phi_n(x) = \langle x|n\rangle$ the corresponding eigenfunction, given by

$$\phi_n(x) = \frac{1}{\sqrt{n!}} \left(-\frac{\sigma}{\sqrt{2}} \frac{d}{dx} + \frac{1}{\sqrt{2}\sigma} x\right)^n \phi_0(x) \tag{7.44}$$

being

$$\phi_0(x) = \frac{1}{\pi^{1/4}\sigma^{1/2}} e^{-x^2/(2\sigma^2)} \tag{7.45}$$

the Gaussian wavefunction of the ground state with energy $E_0 = \hbar\omega/2$. The lowest energies and corresponding eigenfunctions of the quantum particle of mass m in a harmonic potential one-dimensional potential of frequency ω are:

$$E_0 = \frac{1}{2}\hbar\omega \quad \leftrightarrow \quad \phi_0(x) = \frac{1}{\pi^{1/4}\sigma^{1/2}} e^{-x^2/(2\sigma^2)} \tag{7.46}$$

$$E_1 = \frac{3}{2}\hbar\omega \quad \leftrightarrow \quad \phi_1(x) = \frac{1}{\pi^{1/4}\sigma^{1/2}} \sqrt{2}\,\frac{x}{\sigma} e^{-x^2/(2\sigma^2)} \tag{7.47}$$

$$E_2 = \frac{5}{2}\hbar\omega \quad \leftrightarrow \quad \phi_2(x) = \frac{1}{\pi^{1/4}\sigma^{1/2}} \frac{1}{\sqrt{2}} \left(\frac{x^2}{\sigma^2} - 1\right)^2 e^{-x^2/(2\sigma^2)}. \tag{7.48}$$

7.2.1 Properties of Number Operator

We now prove that the eigenvalues n of the number operator $\hat{N} = \hat{a}^+\hat{a}$ are non negative. By using the Dirac notation the eigenvalue equation of $\hat{N}$ reads

$$\hat{a}^+\hat{a}|n\rangle = n|n\rangle . \tag{7.49}$$

We can then write

$$\langle n|\hat{a}^+\hat{a}|n\rangle = n\langle n|n\rangle = n , \tag{7.50}$$

because the eigenstate $|n\rangle$ is normalized to one. On the other hand, we have also

$$\langle n|\hat{a}^+\hat{a}|n\rangle = (\hat{a}|n\rangle)^+(\hat{a}|n\rangle) = |(\hat{a}|n\rangle)|^2 \tag{7.51}$$

Consequently, we get

$$n = |(\hat{a}|n\rangle)|^2 \geq 0 . \tag{7.52}$$

We now show that that if $|n\rangle$ is an eigenstate of the number operator $\hat{N} = \hat{a}^+\hat{a}$ with eigenvalue n then $a|n\rangle$ is an eigenstate of $\hat{N}$ with eigenvalue $n-1$ and $a^+|n\rangle$ is an eigenstate of $\hat{N}$ with eigenvalue $n+1$. In fact, we have

$$\hat{N}\hat{a}|n\rangle = (\hat{a}^+\hat{a})\hat{a}|n\rangle . \tag{7.53}$$

The commutation relation

$$\hat{a}\hat{a}^+ - \hat{a}^+\hat{a} = 1 \tag{7.54}$$

between $\hat{a}$ and $\hat{a}^+$ can be rewritten as

$$\hat{a}^+\hat{a} = \hat{a}\hat{a}^+ - 1 . \tag{7.55}$$

This implies that

$$\hat{N}\hat{a}|n\rangle = (\hat{a}\hat{a}^+ - 1)\hat{a}|n\rangle = \hat{a}(\hat{a}^+\hat{a} - 1)|n\rangle = \hat{a}(\hat{N} - 1)|n\rangle = (n-1)\hat{a}|n\rangle . \tag{7.56}$$

Finally, we obtain

$$\hat{N}\hat{a}^{+}|n\rangle = (\hat{a}^{+}\hat{a})\hat{a}^{+}|n\rangle = \hat{a}^{+}(\hat{a}\hat{a}^{+})|n\rangle = \hat{a}^{+}(\hat{a}^{+}\hat{a}+1)|n\rangle = \hat{a}^{+}(\hat{N}+1)|n\rangle = (n+1)\hat{a}^{+}|n\rangle . \tag{7.57}$$

Taking into account the previous results of this subsection, we can also show that the spectrum of the number operator is the set of integer numbers. We have seen that $\hat{N}$ has a non negative spectrum. This means that $\hat{N}$ possesses a lowest eigenvalue n_0 with $|n_0\rangle$ its eigenstate. This eigenstate $|n_0\rangle$ is such that

$$\hat{a}|n_0\rangle = 0 . \tag{7.58}$$

In fact, on the basis of the results of the previous problem, $\hat{a}|n_0\rangle$ should be eigenstate of $\hat{N}$ with eigenvalue $n_0 - 1$ but this is not possible because n_0 is the lowest eigenvalue of $\hat{N}$. Consequently the state $\hat{a}|n_0\rangle$ is not a good quantum state and we set it equal to 0. In addition, due to the fact that

$$\hat{N}|n_0\rangle = n_0|n_0\rangle = \hat{a}^{+}\hat{a}|n_0\rangle = \hat{a}^{+}\left(\hat{a}|n_0\rangle\right) = \hat{a}^{+}(0) = 0 \tag{7.59}$$

we find that $n_0 = 0$. Thus, the state $|0\rangle$, called vacuum state, is the eigenstate of $\hat{N}$ with eigenvalue 0, i.e.

$$\hat{N}|0\rangle = 0|0\rangle = 0 , \tag{7.60}$$

but also

$$\hat{a}|0\rangle = 0 . \tag{7.61}$$

Due to this equation, it follows that the eigenstates of $\hat{N}$ are only those generated by applying n times the operator $\hat{a}^{+}$ on the vacuum state $|0\rangle$, namely

$$|n\rangle = \frac{1}{\sqrt{n!}}(\hat{a}^{+})^{n}|0\rangle , \tag{7.62}$$

where the factorial is due to the normalization. Finally, we notice that it has been shown in the previous problem that the state $|n\rangle$ has integer eigenvalue n.

7.3 One-Dimensional Scattering

In this section we discuss an example where a particle is scattering with a potential $U(x)$. In particular, we analyze a stationary problem where $U(x) = 0$ for $x < 0$. Thus, for $x \neq 0$ we have a free particle described by the plane wave with unbound energy $E = \hbar^2 k^2/(2m)$. Instead for $x > 0$ we assume that $U(x) \neq 0$ and this implies a much more complicated wavefunction $\psi(x)$ in the region.

We consider the step potential (Fig. 7.3)

Fig. 7.3 Step potential $U(x)$ with $U_0 = 1$

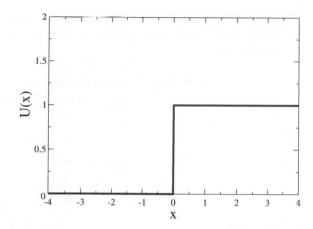

$$U(x) = \begin{cases} 0 & \text{for } x < 0 \\ U_0 & \text{for } x \geq 0 \end{cases} \tag{7.63}$$

The wavefunction is then given by

$$\psi_1(x) = A e^{i k_1 x} + B e^{-i k_1 x} \qquad \text{for } x < 0 \tag{7.64}$$
$$\psi_2(x) = C e^{i k_2 x} \qquad\qquad \text{for } x \geq 0 \tag{7.65}$$

where

$$k_1 = \sqrt{\frac{2m E}{\hbar^2}} \tag{7.66}$$

$$k_2 = \sqrt{\frac{2 m (E - U_0)}{\hbar^2}} \; . \tag{7.67}$$

The coefficients A, B, C have a clear physical meaning: A is the amplitude probability of the incoming wave, B is the amplitude probability of the reflecting wave, and C is the amplitude probability of the transmitted wave. Obviously, both k_1 and k_2 are real numbers if the energy E of the incoming particle is larger than the energy barrier U_0. Instead, if $E < U_0$ then k_2 is purely imaginary, and such that

$$k_2 = i\sqrt{\frac{2 m (U_0 - E)}{\hbar^2}} \; , \tag{7.68}$$

implying an exponential decay of $\psi_2(x)$ in space, i.e.

$$\psi_2(x) = C e^{-\sqrt{\frac{2 m (U_0 - E)}{\hbar^2}} \, x} \; . \tag{7.69}$$

The coefficients A, B, C can be found imposing the boundary conditions of the wavefunction at $x = 0$. The wavefunction and its derivative must be continuous everywhere, i.e.

$$\psi_1(0) = \psi_2(0) \tag{7.70}$$

$$\psi_1'(0) = \psi_2'(0), \tag{7.71}$$

obtaining

$$A + B = C \tag{7.72}$$

$$A - B = \frac{k_2}{k_1}C, \tag{7.73}$$

or equivalently

$$\frac{B}{A} = \frac{k_1 - k_2}{k_1 + k_2} \quad \frac{C}{A} = \frac{2k_1}{k_1 + k_2}. \tag{7.74}$$

Introducing the reflection coefficient

$$\mathcal{R} = \left| \frac{B}{A} \right|^2 \tag{7.75}$$

and the transmission coefficient

$$\mathcal{T} = \frac{k_2}{k_1} \left| \frac{C}{A} \right|^2 = 1 - \mathcal{R} \tag{7.76}$$

one gets

$$\mathcal{T} = 1 - \left(\frac{1 - \sqrt{1 - \frac{U_0}{E}}}{1 + \sqrt{1 - \frac{U_0}{E}}} \right)^2 \tag{7.77}$$

for $E \geq U_0$, and $\mathcal{T} = 0$ for $E < U_0$. Therefore, if the kinetic energy E of a quantum particle is below the energy U_0 of the barrier the transmission probability is zero because the wavefunction decays exponentially (and asymptotically goes to zero) in the classically forbidden region $x \geq 0$.

7.4   One-Dimensional Double-Well Potential

Let us consider a one-dimensional system with classical Hamiltonian

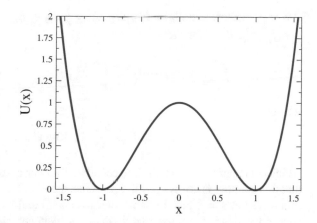

Fig. 7.4 Double-well potential $U(x) = (x^2 - 1)^2$. In this case the maximum U_0 of the energy barrier is $U_0 = 1$

$$H = \frac{p^2}{2m} + U(x),$$

(7.78)

where $U(-x) = U(x)$ is a symmetric double-well potential. In Fig. 7.4 we plot a simple example of symmetric double-well potential. In classical mechanics, if the energy E of the particle is below the maximum of the barrier U_0, there is a forbidden unaccessible region for $|x| < a$, where a is such that $U(a) = U(-a) = E$.

On the contrary, in quantum mechanics the particle can stay also in the classically forbidden region. This phenomenon is called *quantum tunneling*. The stationary Schrödinger equation of the quantum system reads

$$\hat{H}\phi(x) = \left(-\frac{\hbar^2}{2m}\frac{d^2}{dx^2} + U(x)\right)\phi(x) = E\phi(x).$$

(7.79)

Let $\phi_1(x)$ and $\phi_2(x)$ be two exact eigenfunctions of the Schrödinger equation

$$\hat{H}\phi_1 = E_1\phi_1 \quad \text{and} \quad \hat{H}\phi_2 = E_2\phi_2,$$

(7.80)

such that $\phi_1(-x) = \phi_1(x)$ and $\phi_2(-x) = -\phi_2(x)$ and $E_1 \simeq E_2$. To calculate the splitting $\Delta E = E_2 - E_1$, we multiply the first equation by $\phi_2(x)$ and the second by $\phi_1(x)$ and then we subtract the two resulting equations. By integrating from 0 to ∞ we find

$$\Delta E = \frac{\hbar^2}{2m}\frac{\phi_1(0)\phi'_2(0) - \phi'_1(0)\phi_2(0)}{\int_0^\infty \phi_1(x)\phi_2(x)dx}.$$

(7.81)

We write the eigenfunctions $\phi_1(x)$ and $\phi_2(x)$ in terms of the right-localized function

$$\phi_R(x) = \frac{1}{\sqrt{2}}(\phi_1(x) + \phi_2(x)).$$

(7.82)

It is easy to show that $E_R = \langle \phi_R | \hat{H} | \phi_R \rangle = \frac{1}{2}(E_1 + E_2)$. Then, with the approximation $\int_R^\infty \phi_R^2 dx \simeq 1$, namely

$$\int_0^\infty \phi_1(x)\phi_2(x)dx = \int_0^\infty \phi_R^2(x)dx - \frac{1}{2} \simeq 1 - \frac{1}{2} = \frac{1}{2}, \qquad (7.83)$$

we get

$$\Delta E \simeq \frac{2\hbar^2}{m} \phi_R(0)\phi'_R(0), \qquad (7.84)$$

which is an approximate formula to calculate the energy splitting. One should observe that this quantity is always positive, because the tail of the right localized eigenfunction $\phi_R(x)$ at $x = 0$ has the same sign for $\phi_R(0)$ and its derivative $\phi'_R(0)$. Another way to see this is to realize that there are no degeneracies in one-dimensional systems, implying that all pairs of almost degenerate states, from the ground state up, are grouped by odd state above the even state.

7.4.1 One-Dimensional Double-Square-Well Potential

As a specific example, we consider the 1D double-square-well potential (Fig. 7.5) given by

$$U(x) = \begin{cases} V_0 & \text{for } |x| \leq a \\ 0 & \text{for } a < |x| \leq b \\ +\infty & \text{for } |x| > b \end{cases} \qquad (7.85)$$

with $V_0 > 0$. By using the quasi-exact right-localized wavefunction

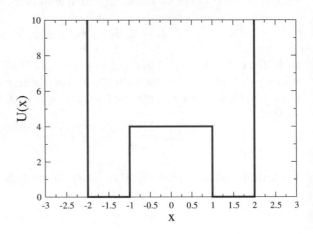

Fig. 7.5 Double-square-well potential $U(x)$ with $a = 1$, $b = 2$, and $U_0 = 4$

$$\phi_R(x) = D\,e^{-\frac{1}{\hbar}\sqrt{2m(V_0-E)}\,x} \tag{7.86}$$

for $0 < x < a$ (classically forbidden region), and

$$\phi_R(x) = A\,e^{\frac{i}{\hbar}\sqrt{2mE}\,x} + B\,e^{-\frac{i}{\hbar}\sqrt{2mE}\,x} \tag{7.87}$$

for $a < x < b$ (classically allowed region), and by imposing matching and normalization conditions we find

$$A = \frac{D}{2}\left(1 - i\sqrt{\frac{V_0-E}{E}}\right)e^{\frac{a}{\hbar}\sqrt{2m(V_0-E)}-\frac{a}{\hbar}\sqrt{E}}, \tag{7.88}$$

$$B = \frac{D}{2}\left(1 + i\sqrt{\frac{V_0-E}{E}}\right)e^{\frac{a}{\hbar}\sqrt{2m(V_0-E)}+\frac{a}{\hbar}\sqrt{E}}, \tag{7.89}$$

and

$$D^2 = \frac{2E}{V_0(b-a)}\,e^{-\frac{2a}{\hbar}\sqrt{2m(V_0-E)}}. \tag{7.90}$$

Finally, we obtain

$$\Delta E \simeq \frac{4\hbar E\sqrt{2m(V_0-E)}}{mV_0(b-a)}\,e^{\frac{2a}{\hbar}\sqrt{2m(V_0-E)}}, \tag{7.91}$$

that is the energy splitting for the double-square-well potential.

The result that we have found is quite general. The analytical formula for the energy splitting can be formally written as

$$\Delta E \simeq A\,e^{-S/\hbar}, \tag{7.92}$$

where A is the tunneling amplitude and S is the tunneling action, which is the classical action

$$S = \int_{-a}^{a} p\,dx = \int_{-a}^{a}\sqrt{2m(V_0-E)}\,dx = 2a\sqrt{2m(V_0-E)} \tag{7.93}$$

inside the classically forbidden region $|x| < a$. In conclusion, the splitting of the energy levels in the double-well potential is strictly related to the quantum tunneling, i.e. to the fact that a quantum particles can tunnel the energy barrier.

7.5   WKB Method

The quantized energy levels of the one-dimensional Schrödinger problem

$$\left[-\frac{\hbar^2}{2m}\frac{\partial^2}{\partial x^2} + U(x)\right]\phi(x) = E\phi(x) ,\tag{7.94}$$

can be obtained by using the so-called WKB method, developed in 1926 by Gregor Wentzel, Hendrik Anthony Kramers, and Leon Brillouin. This method is based on the observation that the wavefunction $\phi(x)$ can be formally written as

$$\phi(x) = e^{\frac{i}{\hbar}\sigma(x)} ,\tag{7.95}$$

where the phase $\sigma(x)$ is a complex function. Inserting this expression into the Schrödinger equation one finds that $\sigma(x)$ satisfies the differential equation

$$(\sigma'(x))^2 + \frac{\hbar}{i}\sigma''(x) = 2m\,(E - U(x)) ,\tag{7.96}$$

where $\sigma' = d\sigma/dx$ and $\sigma'' = d^2\sigma/dx^2$. The WKB expansion for the phase $\sigma(x)$ is given by

$$\sigma(x) = \sum_{k=0}^{+\infty}\left(\frac{\hbar}{i}\right)^k \sigma_k(x) .\tag{7.97}$$

Substituting (7.97) into (7.96) and comparing like powers of $\hbar$ gives the recursion relation $(n > 0)$

$$\sigma''_{n-1} + \sum_{k=0}^{n}\sigma'_k\,\sigma'_{n-k} = 0 \tag{7.98}$$

with initial condition

$$(\sigma'_0)^2 = 2m\,(E - U(x)) .\tag{7.99}$$

The quantization condition is obtained by requiring the single-valuedness of the wavefunction, namely

$$\oint d\sigma = \sum_{k=0}^{+\infty}\left(\frac{\hbar}{i}\right)^k \oint d\sigma_k = 2\pi\hbar n ,\tag{7.100}$$

where $n \in \mathbb{N}$ is a quantum number. The zero order term is given by

$$\oint d\sigma_0 = \oint \sqrt{2m(E - U(x))}\,dx = \oint p(x)\,dx \tag{7.101}$$

where $p(x) = \sqrt{2m(E - U(x))}$ is the linear momentum. Thus, the zero order of the WKB $\hbar$ expansion gives

$$\oint p(x) \, dx = 2 \int \sqrt{2m(E - U(x))} \, dx = h \, n \,, \qquad (7.102)$$

that is nothing else than the Bohr-Sommerfeld-Wilson quantization rule of old quantum mechanics, with $h = 2\pi\hbar$. Instead, the first odd term gives the Maslov correction

$$\left(\frac{\hbar}{i}\right) \oint d\sigma_1 = \left(\frac{\hbar}{i}\right) \frac{1}{4} \ln p|_{contour} = -\pi\hbar \,. \qquad (7.103)$$

In this way, to the first order of the WKB $\hbar$ expansion one obtains

$$\oint p(x) \, dx = h \left(n + \frac{1}{2}\right) \,. \qquad (7.104)$$

Obviously, there are corrections to this formula. Remarkably, all the other odd terms vanish when integrated along the closed contour because they are exact differentials. So the quantization condition (7.100) can be written

$$\sum_{k=0}^{+\infty} (\frac{\hbar}{i})^{2k} \oint d\sigma_{2k} = h \left(n + \frac{1}{2}\right) \,, \qquad (7.105)$$

thus a sum over even-numbered terms only. For instance, the next non-zero term is

$$(\frac{\hbar}{i})^2 \oint d\sigma_2 = -\frac{\hbar^2}{12} \frac{\partial^2}{\partial E^2} \int \frac{U'(x)^2}{\sqrt{2m(E - U(x))}} \, dx \,. \qquad (7.106)$$

Higher-order corrections quickly increase in complexity but in specific cases they can be calculated. For some special potentials $U(x)$ one determines analytically all orders of the WKB expansion, showing that the series is convergent and gives precisely the exact result.

7.6 Three-Dimensional Separable Potential

The stationary Schrödinger equation for a particle moving in three dimensions is given by

$$\hat{H} \, \phi(x, y, z) = E \, \phi(x, y, z) \,, \qquad (7.107)$$

where

$$\hat{H} = -\frac{\hbar^2}{2m} \left(\frac{\partial^2}{\partial x^2} + \frac{\partial^2}{\partial y^2} + \frac{\partial^2}{\partial z^2}\right) + U(x, y, z) \qquad (7.108)$$

is the Hamiltonian operator of this three-dimensional (3D) problem.

We assume that the external potential $U(x, y, z)$ is separable, i.e. of type

$$U(x, y, z) = U^{(1)}(x) + U^{(2)}(y) + U^{(3)}(z), \tag{7.109}$$

where $U^{(1)}(x)$ is a potential acting only on the coordinate x, $U^{(2)}(y)$ is a potential acting on the coordinate y only, and $U^{(3)}(z)$ is a potential acting on the z-coordinate only. The problem under consideration is described by the stationary Schrödinger equation

$$-\frac{\hbar^2}{2m} \left(\frac{\partial^2}{\partial x^2} + \frac{\partial^2}{\partial y^2} + \frac{\partial^2}{\partial z^2} \right) \phi(x, y, z) + \left(U^{(1)}(x) + U^{(2)}(y) + U^{(3)}(z) \right) \phi(x, y, z) = E \phi(x, y, z).$$
$$\tag{7.110}$$

The fact that the potential is separable suggests to us the following factorization for the wavefunction

$$\phi(x, y, z) = \phi^{(1)}(x) \, \phi^{(2)}(y) \, \phi^{(3)}(z), \tag{7.111}$$

where $\phi^{(1)}(x)$ is a function of coordinate x alone, $\phi^{(2)}(y)$ is a function of the coordinate y only, and $\phi^{(3)}(z)$ is a function of the coordinate z only. We also assume that

$$E = E^{(1)} + E^{(2)} + E^{(3)}, \tag{7.112}$$

where $E^{(1)}$ is the energy of motion along the x coordinate, $E^{(2)}$ is the energy of the motion along the y-coordinate, and $E^{(3)}$ is the energy of the motion along the z-coordinate. The Schrödinger equation becomes

$$\phi^{(2)}(y) \, \phi^{(3)}(z) \left(-\frac{\hbar^2}{2m} \frac{\partial^2}{\partial x^2} + U^{(1)}(x) - E^{(1)} \right) \phi^{(1)}(x)$$
$$+ \phi^{(1)}(x) \, \phi^{(3)}(z) \left(-\frac{\hbar^2}{2m} \frac{\partial^2}{\partial y^2} + U^{(2)}(y) - E^{(2)} \right) \phi^{(2)}(y)$$
$$+ \phi^{(1)}(x) \, \phi^{(2)}(y) \left(-\frac{\hbar^2}{2m} \frac{\partial^2}{\partial z^2} + U^{(3)}(z) - E^{(3)} \right) \phi^{(3)}(z) = 0. \tag{7.113}$$

Dividing by $\phi^{(1)}(x) \, \phi^{(2)}(y) \, \phi^{(3)}(z)$ yields

$$\frac{1}{\phi^{(1)}(x)} \left(-\frac{\hbar^2}{2m} \frac{\partial^2}{\partial x^2} + U^{(1)}(x) - E^{(1)} \right) \phi^{(1)}(x)$$
$$+ \frac{1}{\phi^{(2)}(y)} \left(-\frac{\hbar^2}{2m} \frac{\partial^2}{\partial y^2} + U^{(2)}(y) - E^{(2)} \right) \phi^{(2)}(y)$$
$$+ \frac{1}{\phi^{(3)}(z)} \left(-\frac{\hbar^2}{2m} \frac{\partial^2}{\partial z^2} + U^{(3)}(z) - E^{(3)} \right) \phi^{(3)}(z) = 0. \tag{7.114}$$

So we have the sum of three terms: one that depends only on the variable x, one that depends only on the variable y, one that depends only on the variable z. And the sum must be equal to zero. For mathematical consistency, each of the three terms must be equal to zero. Thus, three independent Schrödinger equations are found:

$$
\left(-\frac{\hbar^2}{2m}\frac{\partial^2}{\partial x^2} + U^{(1)}(x)\right)\phi^{(1)}(x) = E^{(1)}\phi^{(1)}(x),
$$

$$
\left(-\frac{\hbar^2}{2m}\frac{\partial^2}{\partial y^2} + U^{(2)}(y)\right)\phi^{(2)}(y) = E^{(2)}\phi^{(2)}(y),
$$

$$
\left(-\frac{\hbar^2}{2m}\frac{\partial^2}{\partial z^2} + U^{(3)}(z)\right)\phi^{(3)}(z) = E^{(3)}\phi^{(3)}(z). \tag{7.115}
$$

In conclusion, if the potential is separable the 3D problem can be leads back to three 1D problems.

7.6.1 Three-Dimensional Harmonic Potential

As an example, let us consider a quantum particle under the action of the three-dimensional harmonic potential

$$
U(x, y, z) = \frac{1}{2}m\omega_1^2 x^2 + \frac{1}{2}m\omega_2^2 y^2 + \frac{1}{2}m\omega_3^2 z^2. \tag{7.116}
$$

The potential is separable, in fact

$$
U(x, y, z) = U^{(1)}(x) + U^{(2)}(y) + U^{(3)}(z), \tag{7.117}
$$

where

$$
U^{(1)}(x) = \frac{1}{2}m\omega_1^2 x^2 \tag{7.118}
$$

$$
U^{(2)}(y) = \frac{1}{2}m\omega_2^2 y^2 \tag{7.119}
$$

$$
U^{(3)}(z) = \frac{1}{2}m\omega_3^2 z^2. \tag{7.120}
$$

Based on what has been discussed above, the 3D problem is reduced to three independent problems of a particle in harmonic potential:

$$
\left(-\frac{\hbar^2}{2m}\nabla^2 + \frac{1}{2}m\omega_1^2 x^2 + \frac{1}{2}m\omega_2^2 y^2 + \frac{1}{2}m\omega_3^2 z^2\right)\phi_{n_1 n_2 n_3}(\mathbf{r}) = E_{n_1 n_2 n_3}\phi_{n_1 n_2 n_3}(\mathbf{r}),
$$

$$
\tag{7.121}
$$

where

$$\phi_{n_1 n_2 n_3}(\mathbf{r}) = \phi_{n_1}^{(1)}(x)\,\phi_{n_2}^{(2)}(y)\,\phi_{n_3}^{(3)}(z) \tag{7.122}$$

and

$$E_{n_1 n_2 n_3} = \hbar\omega_1\left(n_1 + \frac{1}{2}\right) + \hbar\omega_2\left(n_2 + \frac{1}{2}\right) + \hbar\omega_3\left(n_3 + \frac{1}{2}\right) \tag{7.123}$$

remembering the quantization formula of the energy levels of the 1D harmonic oscillator. Here n_1, n_2, n_3 are three natural numbers: the three quantum numbers that characterize the energy levels.

In the particular case $\omega_1 = \omega_2 = \omega_3$, indicating by ω the three equal frequencies, we have

$$\left(-\frac{\hbar^2}{2m}\nabla^2 + \frac{1}{2}m\omega^2(x^2 + y^2 + z^2)\right)\phi_{n_1 n_2 n_3}(\mathbf{r}) = E_{n_1 n_2 n_3}\phi_{n_1 n_2 n_3}(\mathbf{r}), \tag{7.124}$$

where

$$\phi_{n_1 n_2 n_3}(\mathbf{r}) = \phi_{n_1}^{(1)}(x)\,\phi_{n_2}^{(2)}(y)\,\phi_{n_3}^{(3)}(z) \tag{7.125}$$

and

$$E_{n_1 n_2 n_3} = \hbar\omega\left(n_1 + n_2 + n_3 + \frac{3}{2}\right). \tag{7.126}$$

The ground state has energy $E_{000} = 3\hbar\omega/2$, while the first excited state has energy $E_{100} = E_{010} = E_{001} = 5\hbar\omega/2$. The first excited state turns out to be degenerate: there are three independent wave functions that produce the same energy.

Further Reading

Solvable problems of quantum mechanics are discussed in detail here:
Cohen-Tannoudji, C., Diu, B., Laloe, F.: Quantum Mechanics: Basic Concepts, Tools, and Applications. Wiley (2019)
Landau, L.D., Lifshitz, E.M.: Quantum Mechanics: Non-Relativistic Theory. Pergamon (1981)
Sakurai, J.J., Napolitano, J.: Modern Quantum Mechanics. Cambridge Univ. Press (2020)
An extremely important historical paper on ladder operators and number operator is:
Dirac, P.A.M.: Proc. R. Soc. Lond. A **114**, 243 (1927)
The WKB method to all orders of the $\hbar$ expansion is discussed in:
Robnik, M., Salasnich, L.: J. Phys. A **30**, 1711 (1997)

Chapter 8
Modern Quantum Physics of Atoms

In this chapter we analyze the quantum mechanics of the hydrogen atom by using the three-dimensional Schrödinger equation. We also introduce the spin of the electron, giving a rigorous theoretical justification in the basis of the Dirac equation. Finally, we discuss how the quantum energy levels of atoms are modified in the presence of an electric field (Stark effect) and of a magnetic field (Zeeman effect).

8.1 Electron in the Hydrogen Atom

The neutral hydrogen atom is formed by an electron orbiting around the nucleus that is formed by a single proton. The stationary Schrödinger equation for an electron of charge $-e$ and mass m_e moving in three dimensions under the action of the Coulomb force exerted by the proton is given by

$$\hat{H}\,\phi(x, y, z) = E\,\phi(x, y, z)\,, \tag{8.1}$$

where

$$\hat{H} = -\frac{\hbar^2}{2m_e}\left(\frac{\partial^2}{\partial x^2} + \frac{\partial^2}{\partial y^2} + \frac{\partial^2}{\partial z^2}\right) + U(x, y, z) \tag{8.2}$$

is the Hamiltonian operator of this three-dimensional (3D) problem and

$$U(x, y, z) = -\frac{e^2}{4\pi\varepsilon_0\sqrt{x^2 + y^2 + z^2}} \tag{8.3}$$

is the potential energy of the electron due to the Coulomb force.

© The Author(s), under exclusive license to Springer Nature Switzerland AG 2022
L. Salasnich, *Modern Physics*, UNITEXT for Physics,
https://doi.org/10.1007/978-3-030-93743-0_8

Unfortunately, in the problem under consideration the potential $U(x, y, z)$ is not separable in Cartesian coordinates. On the other hand, if we introduce spherical polar coordinates, of radius r and angles ϕ and θ such that

$$x = r \, \cos(\phi) \, \sin(\theta) \tag{8.4}$$

$$y = r \, \sin(\phi) \, \sin(\theta) \tag{8.5}$$

$$z = r \, \cos(\theta) \tag{8.6}$$

the Coulomb potential is separable in spherical polar coordinates because it depends only on r, i.e.

$$U(r) = -\frac{e^2}{4\pi\varepsilon_0 r} \, . \tag{8.7}$$

The Laplace operator

$$\nabla^2 = \frac{\partial^2}{\partial x^2} + \frac{\partial^2}{\partial y^2} + \frac{\partial^2}{\partial z^2} \tag{8.8}$$

expressed in spherical polar coordinates is given by

$$\nabla^2 = \nabla_r^2 + \frac{\hat{\mathcal{L}}_{\phi,\theta}^2}{r^2} \tag{8.9}$$

where

$$\nabla_r^2 = \frac{\partial^2}{\partial r^2} + \frac{1}{r}\frac{\partial}{\partial r} \tag{8.10}$$

while

$$\hat{\mathcal{L}}_{\phi,\theta}^2 = \frac{1}{\sin^2(\theta)}\frac{\partial^2}{\partial\phi^2} + \frac{1}{\sin(\theta)}\frac{\partial}{\partial\theta}\left(\sin(\theta)\frac{\partial}{\partial\theta}\right) . \tag{8.11}$$

In spherical polar coordinates the Hamiltonian operator of the electron in the hydrogen atom is therefore given by

$$\hat{H} = -\frac{\hbar^2}{2m_e}\nabla_r^2 + \frac{\hat{L}^2}{2m_e r^2} - \frac{e^2}{4\pi\varepsilon_0 r} , \tag{8.12}$$

where

$$\hat{L}^2 = -\hbar^2\hat{\mathcal{L}}_{\phi,\theta}^2 \tag{8.13}$$

is called quantum operator of the square of the orbital angular momentum. This name is fully motivated by the fact that, given the square of the orbital angular momentum L of classical mechanics, i.e.

$$L^2 = (\mathbf{r} \wedge \mathbf{p})^2 \tag{8.14}$$

and applying the quantization rule $\mathbf{p} \leftrightarrow -i\hbar\nabla$, we get just $\hat{L}^2$ if spherical polar coordinates are used.

8.1.1 Schrödinger Equation in Spherical Polar Coordinates

Therefore, in spherical polar coordinates, the Schrödinger equation of the electron in the hydrogen atom is given by

$$\left(-\frac{\hbar^2}{2m_e}\nabla_r^2 + \frac{\hat{L}^2}{2m_e r^2} - \frac{e^2}{4\pi\varepsilon_0 r}\right)\phi(r, \phi, \theta) = E\,\phi(r, \phi, \theta)\,, \qquad (8.15)$$

where the differential operator ∇_r^2 involves only the derivatives of the radial coordinate r, while the differential operator $\hat{L}^2$ involves only the derivatives of the angular coordinates ϕ and θ.

In the problem we can then impose the following factorization for the wave function

$$\phi(r, \theta, \phi) = R(r)\,Y(\theta, \phi)\,. \qquad (8.16)$$

It is possible to prove that the operator $\hat{L}^2$ satisfies the eigenvalue equation

$$\hat{L}^2 Y_{lm_l}(\theta, \phi) = \hbar^2 l(l+1)\,Y_{lm_l}(\theta, \phi) \qquad (8.17)$$

where l is a natural number called quantum number of the orbital angular momentum while m_l is an integer number called magnetic quantum number and such that

$$m_l = -l, -l+1, -l+2, \ldots, l-2, l-1, l\,. \qquad (8.18)$$

Then for every value of l there are $2l + 1$ possible values for m_l. The functions $Y_{lm_l}(\theta, \phi)$ are called spherical harmonics. For spherical harmonics there is another important eigenvalue equation

$$\hat{L}_z Y_{lm_l}(\theta, \phi) = \hbar m_l\,Y_{lm_l}(\theta, \phi) \qquad (8.19)$$

involving the differential operator

$$\hat{L}_z = -i\hbar\frac{\partial}{\partial\phi} \qquad (8.20)$$

associated with the third component of the orbital angular momentum.

Finally, it is possible to show that

$$\left(-\frac{\hbar^2}{2m_e}\nabla_r^2 + \frac{\hbar^2 l(l+1)}{2m_e r^2} - \frac{e^2}{4\pi\varepsilon_0 r} \right) R_{nl}(r) = -\frac{m_e e^4}{32\pi^2\varepsilon_0^2\hbar^2}\frac{1}{n^2} R_{nl}(r), \quad (8.21)$$

where n is a natural number different from zero, called principal quantum number such that, once n is fixed, l can assume the following values $l = 0, 1, 2, \ldots, n-1$.

In conclusion, we can write for the electron in the hydrogen atom the stationary Schrödinger equation

$$\hat{H}\phi_{nlm_l}(r, \theta, \phi) = E_n \, \phi_{nlm_l} \qquad (8.22)$$

with eigenfunctions

$$\phi_{nlm_l}(r, \theta, \phi) = R_{nl}(r) \, Y_{lm_l}(\theta, \phi) \qquad (8.23)$$

and eigenvalues

$$E_n = -\frac{m_e e^4}{32\pi^2\varepsilon_0^2\hbar^2}\frac{1}{n^2} = -\frac{m_e e^4}{8\varepsilon_0^2 h^2}\frac{1}{n^2} = -\frac{13.6\,\text{eV}}{n^2}. \qquad (8.24)$$

This last one is exactly the same formula of quantization of energy levels first obtained by Bohr in 1913.

The formalism of the old Bohr's quantum mechanics does not provide information about the wave function of the electron. Instead, Schrödinger formalism provides this information as well. In fact, in the formula

$$\phi_{nlm_l}(r, \theta, \phi) = R_{nl}(r) \, Y_{lm_l}(\theta, \phi) \qquad (8.25)$$

both the radial functions $R_{nl}(r)$ and the spherical harmonics $Y_{lm_l}(\theta, \phi)$ can be computed explicitly, although the calculation is not at all straightforward. For example, one can show that for the ground state we have

$$R_{10}(r) = \frac{2}{r_0^{3/2}} e^{-r/r_0}, \qquad (8.26)$$

$$Y_{00}(\theta, \phi) = \frac{1}{\sqrt{4\pi}}, \qquad (8.27)$$

with $r_0 = 4\pi\varepsilon_0\hbar^2/(m_e e^2) = 0.53 \cdot 10^{-10}$ m the so-called Bohr radius.

8.1.2 Selection Rules

Using the ket formalism of Dirac we can write, for the electron in the hydrogen atom, the stationary Schrödinger equation

$$\hat{H}|n\,l\,m_l\rangle = E_n\,|n\,l\,m_l\rangle \qquad (8.28)$$

with

$$E_n = -\frac{13.6 \text{ eV}}{n^2}.$$ (8.29)

In addition, the following equations hold

$$\hat{L}^2 |n\, l\, m_l\rangle = \hbar^2 l(l+1) |n\, l\, m_l\rangle,$$ (8.30)
$$\hat{L}_z |n\, l\, m_l\rangle = \hbar m_l |n\, l\, m_l\rangle,$$ (8.31)

where the three quantum numbers n, l, m_l are such that:

$$n = 1, 2, 3, \ldots,$$

$$l = 0, 1, 2, \ldots, n-1,$$

$$m_l = -l, -l+1, -l+2, \ldots, l-2, l-1, l.$$

Considering the Schrödinger equation for the electron in the atom of hydrogen and the interaction of the electron with the electromagnetic field, and therefore the interaction of the electron with the photon, it is possible to show (within the so-called dipole approximation) that the transition electromagnetic between the state $|n'\, l'\, m_l'\rangle$ and the state $|n'\, l\, m_l\rangle$ occurs only if the following selection rules are met

$$\Delta l = \pm 1 \qquad \Delta m_l = 0, \pm 1,$$ (8.32)

where $\Delta l = l' - l$ and $\Delta m_l = m_l' - m_l$. These selection rules are confirmed by experimental evidence. So, once again, modern quantum mechanics, based on the Schrödinger equation, shows a predictive power considerably higher than the old quantum mechanics of Bohr, Wilson and Sommerfeld.

8.2 Pauli Exclusion Principle and the Spin

The Schrödinger equation for the electron describes extremely well the quantum properties of the hydrogen atom: its energy spectrum and the allowed electromagnetic transitions. Applied to explain many-electron atoms Schrödinger equation continues to work very well if one takes into account a principle, called *exclusion principle*, formulated by Wolfang Pauli in 1925: "There cannot be two electrons in the same single-particle quantum state". To explain Mendeleev's periodic table of elements, Pauli used his exclusion principle assuming also that the electron has an intrinsic angular momentum, called *spin*, characterized by only two measurable values:

$$-\frac{\hbar}{2} \quad \text{and} \quad \frac{\hbar}{2}.$$

In 1928 Paul Dirac generalized the Schrödinger equation to the relativistic case. Among the consequences of this generalized equation, known as the Dirac equation, there is just the spin of the electron. From Dirac equation, and other subsequent experimental and theoretical studies, it is deduced in fact that every quantum particle is characterized by an intrinsic angular momentum

$$\hat{\mathbf{S}} = (\hat{S}_x, \hat{S}_y, \hat{S}_z) \tag{8.33}$$

called spin. This spin is a vector whose three components are operators such that the square of the spin $\hat{S}^2$ and the third component $\hat{S}_z$ of the spin satisfy the following eigenvalue equations

$$\hat{S}^2 |s\, m_s\rangle = \hbar^2\, s(s+1)|s\, m_s\rangle\,, \tag{8.34}$$

$$\hat{S}_z |s\, m_s\rangle = \hbar m_s |s\, m_s\rangle\,. \tag{8.35}$$

The numbers s and m_s are called respectively spin quantum number and quantum number of the third spin component. Fixed s, the number m_s can assume $2s + 1$ values, given by $m_s = -s, -s + 1, -s + 2, \ldots, s - 2, s - 1, s$. In the case of the electron, from the Dirac equation it is found in particular that

$$s = \frac{1}{2} \quad \text{and consequently} \quad m_s = -\frac{1}{2}, \frac{1}{2}, \tag{8.36}$$

and, as conjectured by Pauli, with the two possible values $-\hbar/2$ and $\hbar/2$ for $\hat{S}_z$. In the case of electrons the quantum number $s = 1/2$ is always the same, while the quantum number m_s can be $-1/2$ or $1/2$. In the particular case of the electron in a stationary state in hydrogen atom we have then that

$$|n\, l\, m_l\, m_s\rangle$$

is Dirac's ket representing all its quantum numbers: n is the principal quantum number, l is the orbital angular momentum quantum number, m_l is the magnetic quantum number (also called of the third component of the orbital orbital angular momentum), m_s is the spin quantum number (more properly the quantum number of the third component of the spin).

8.2.1 Semi-integer and Integer Spin: Fermions and Bosons

As already said, not only the electron, but any particle is characterized by a spin quantum number s. The particles that have a spin quantum number s semi-integer positive are called fermionic particles (or fermions). Particles that have a positive integer spin s quantum number are called bosonic particles (or bosons). Clearly the electron is a fermionic particle with $s = 1/2$, as well as the proton and neutron. Experiments show instead the photon is a bosonic particle with $s = 1$. The spin of a non elementary particle is the sum of the spin of the elementary particles that constitute this non-elementary particle.

8.3 The Dirac Equation

The classical energy of a nonrelativistic free particle is given by

$$E = \frac{\mathbf{p}^2}{2m} , \tag{8.37}$$

where $\mathbf{p}$ is the linear momentum and m the mass of the particle. We have seen that the Schrödinger equation of the corresponding quantum particle with wavefunction $\psi(\mathbf{r}, t)$ is easily obtained by imposing the quantization prescription

$$E \to i\hbar\frac{\partial}{\partial t} , \qquad \mathbf{p} \to -i\hbar\nabla . \tag{8.38}$$

In this way one gets the time-dependent Schrödinger equation of the free particle, namely

$$i\hbar\frac{\partial}{\partial t}\psi(\mathbf{r}, t) = -\frac{\hbar^2}{2m}\nabla^2\psi(\mathbf{r}, t) , \tag{8.39}$$

obtained for the first time in 1926 by Erwin Schrödinger. The classical energy of a relativistic free particle is instead given by

$$E = \sqrt{\mathbf{p}^2c^2 + m^2c^4} , \tag{8.40}$$

where c is the speed of light in the vacuum. By applying directly the quantization prescription (8.38) one finds

$$i\hbar\frac{\partial}{\partial t}\psi(\mathbf{r}, t) = \sqrt{-\hbar^2c^2\nabla^2 + m^2c^4}\,\psi(\mathbf{r}, t) . \tag{8.41}$$

This equation is quite suggestive but the square-root operator on the right side is a very difficult mathematical object. For this reason in 1927 Oskar Klein and Walter Gordon suggested to start with

$$E^2 = \mathbf{p}^2c^2 + m^2c^4 \tag{8.42}$$

and then to apply the quantization prescription (8.38). In this way one obtains

$$-\hbar^2 \frac{\partial^2}{\partial t^2} \psi(\mathbf{r}, t) = \left(-\hbar^2 c^2 \nabla^2 + m^2 c^4\right) \psi(\mathbf{r}, t) \tag{8.43}$$

the so-called Klein-Gordon equation, which can be re-written in the following form

$$\left(\frac{1}{c^2} \frac{\partial^2}{\partial t^2} - \nabla^2 + \frac{m^2 c^2}{\hbar^2}\right) \psi(\mathbf{r}, t) = 0, \tag{8.44}$$

i.e. a generalization of Maxwell's wave equation for massive particles. This equation has two problems: (i) it admits solutions with negative energy; (ii) the space integral over the entire space of the non negative probability density $\rho(\mathbf{r}, t) = |\psi(\mathbf{r}, t)|^2 \geq 0$ is generally not time-independent, namely

$$\frac{d}{dt} \int_{\mathbb{R}^3} \rho(\mathbf{r}, t) \, d^3\mathbf{r} \neq 0. \tag{8.45}$$

Nowadays we know that to solve completely these two problems it is necessary to promote $\psi(\mathbf{r}, t)$ to a quantum field operator. Within this second-quantization (quantum field theory) approach the Klein-Gordon equation is now used to describe relativistic particles with spin zero, like the pions or the Higgs boson.

In 1928 Paul Dirac proposed a different approach to the quantization of the relativistic particle. To solve the problem of Eq. (8.45) he considered a wave equation with only first derivatives with respect to time and space and introduced the classical energy

$$E = c \, \hat{\boldsymbol{\alpha}} \cdot \mathbf{p} + \hat{\beta} mc^2, \tag{8.46}$$

such that squaring it one recovers Eq. (8.42). This condition is fulfilled only if $\hat{\boldsymbol{\alpha}} = (\hat{\alpha}_1, \hat{\alpha}_2, \hat{\alpha}_3)$ and $\hat{\beta}$ satisfy the following algebra of matrices

$$\hat{\alpha}_1^2 = \hat{\alpha}_2^2 = \hat{\alpha}_3^2 = \hat{\beta}^2 = \hat{I}, \tag{8.47}$$

$$\hat{\alpha}_i \hat{\alpha}_j + \hat{\alpha}_j \hat{\alpha}_i = \hat{0}, \quad i \neq j \tag{8.48}$$

$$\hat{\alpha}_i \hat{\beta} + \hat{\beta} \hat{\alpha}_i = \hat{0}, \quad \forall i \tag{8.49}$$

where $\hat{I}$ is the identity matrix and $\hat{0}$ is the zero matrix. The smallest dimension in which the so-called Dirac matrices $\hat{\alpha}_i$ and $\hat{\beta}$ can be realized is four. In particular, one can write

$$\hat{\alpha}_i = \begin{pmatrix} \hat{0}_2 & \hat{\sigma}_i \\ \hat{\sigma}_i & \hat{0}_2 \end{pmatrix}, \qquad \hat{\beta} = \begin{pmatrix} \hat{I}_2 & \hat{0}_2 \\ \hat{0}_2 & -\hat{I}_2 \end{pmatrix}, \tag{8.50}$$

where $\hat{I}_2$ is the 2×2 identity matrix, $\hat{0}_2$ is the 2×2 zero matrix, and

$$\hat{\sigma}_1 = \begin{pmatrix} 0 & 1 \\ 1 & 0 \end{pmatrix} \qquad \hat{\sigma}_2 = \begin{pmatrix} 0 & -i \\ i & 0 \end{pmatrix} \qquad \hat{\sigma}_3 = \begin{pmatrix} 1 & 0 \\ 0 & -1 \end{pmatrix} \tag{8.51}$$

are the Pauli matrices. Eq. (8.46) with the quantization prescription (8.38) gives

$$i\hbar\frac{\partial}{\partial t}\Psi(\mathbf{r},t)=\left(-i\hbar c\,\hat{\boldsymbol{\alpha}}\cdot\nabla+\hat{\beta}mc^2\right)\Psi(\mathbf{r},t)\,,\tag{8.52}$$

which is the Dirac equation for a free particle. Notice that the wavefunction $\Psi(\mathbf{r},t)$ has four components in the abstract space of Dirac matrices, i.e. this spinor field can be written

$$\Psi(\mathbf{r},t)=\begin{pmatrix}\psi_1(\mathbf{r},t)\\\psi_2(\mathbf{r},t)\\\psi_3(\mathbf{r},t)\\\psi_4(\mathbf{r},t)\end{pmatrix}.\tag{8.53}$$

In explicit matrix form the Dirac equation is thus given by

$$i\hbar\frac{\partial}{\partial t}\begin{pmatrix}\psi_1(\mathbf{r},t)\\\psi_2(\mathbf{r},t)\\\psi_3(\mathbf{r},t)\\\psi_4(\mathbf{r},t)\end{pmatrix}=\hat{H}\begin{pmatrix}\psi_1(\mathbf{r},t)\\\psi_2(\mathbf{r},t)\\\psi_3(\mathbf{r},t)\\\psi_4(\mathbf{r},t)\end{pmatrix}\tag{8.54}$$

where

$$\hat{H}=\begin{pmatrix}mc^2 & 0 & -i\hbar c\frac{\partial}{\partial z} & -i\hbar c(\frac{\partial}{\partial x}-i\frac{\partial}{\partial y})\\0 & mc^2 & -i\hbar c(\frac{\partial}{\partial x}+i\frac{\partial}{\partial y}) & i\hbar c\frac{\partial}{\partial z}\\-i\hbar c\frac{\partial}{\partial z} & -i\hbar c(\frac{\partial}{\partial x}-i\frac{\partial}{\partial y}) & -mc^2 & 0\\-i\hbar c(\frac{\partial}{\partial x}+i\frac{\partial}{\partial y}) & i\hbar c\frac{\partial}{\partial z} & 0 & -mc^2\end{pmatrix}.\tag{8.55}$$

It is easy to show that the Dirac equation satisfies the differential law of current conservation. In fact, left-multiplying Eq. (8.52) by

$$\Psi^+(\mathbf{r},t)=\left(\psi_1^*(\mathbf{r},t),\psi_2^*(\mathbf{r},t),\psi_3^*(\mathbf{r},t),\psi_4^*(\mathbf{r},t)\right)\tag{8.56}$$

we get

$$i\hbar\Psi^+\frac{\partial\Psi}{\partial t}=-i\hbar c\,\Psi^+\hat{\boldsymbol{\alpha}}\cdot\nabla\Psi+mc^2\Psi^+\hat{\beta}\Psi\,.\tag{8.57}$$

Considering the Hermitian conjugate of the Dirac equation (8.52) and right-multiplying it by $\Psi(\mathbf{r},t)$ we find instead

$$-i\hbar\frac{\partial\Psi^+}{\partial t}\Psi=i\hbar c\,\hat{\boldsymbol{\alpha}}\cdot\nabla\Psi^+\Psi+mc^2\Psi^+\hat{\beta}\Psi\,.\tag{8.58}$$

Subtracting the last two equations we obtain the continuity equation

$$\frac{\partial}{\partial t}\rho(\mathbf{r}, t) + \nabla \cdot \mathbf{j}(\mathbf{r}, t) = 0 \,, \tag{8.59}$$

where

$$\rho(\mathbf{r}, t) = \Psi^+(\mathbf{r}, t)\Psi(\mathbf{r}, t) = \sum_{i=1}^{4} |\psi_i(\mathbf{r}, t)|^2 \tag{8.60}$$

is the probability density, and $\mathbf{j}(\mathbf{r}, t)$ is the probability current with three components

$$j_k(\mathbf{r}, t) = c\, \Psi^+(\mathbf{r}, t)\hat{\alpha}_k \Psi(\mathbf{r}, t) \,. \tag{8.61}$$

Finally, we observe that from the continuity equation (8.59) one finds

$$\frac{d}{dt} \int_{\mathbb{R}^3} \rho(\mathbf{r}, t)\, d^3\mathbf{r} = 0 \,, \tag{8.62}$$

by using the divergence theorem and imposing a vanishing current density on the border at infinity. Thus, contrary to the Klein-Gordon equation, the Dirac equation does not have the probability density problem.

8.3.1 The Pauli Equation and the Spin

In this subsection we analyze the non-relativistic limit of the Dirac equation. Let us suppose that the relativistic particle has the electric charge q. In presence of an electromagnetic field, by using the Gauge-invariant substitution

$$i\hbar\frac{\partial}{\partial t} \to i\hbar\frac{\partial}{\partial t} - q\,\phi(\mathbf{r}, t) \tag{8.63}$$

$$-i\hbar\nabla \to -i\hbar\nabla - q\mathbf{A}(\mathbf{r}, t) \tag{8.64}$$

in Eq. (8.52), we obtain

$$i\hbar\frac{\partial}{\partial t}\Psi(\mathbf{r}, t) = \left(c\,\hat{\boldsymbol{\alpha}} \cdot \left(\hat{\mathbf{p}} - q\mathbf{A}(\mathbf{r}, t)\right) + \hat{\beta}\, mc^2 + q\,\phi(\mathbf{r}, t)\right)\Psi(\mathbf{r}, t) \,, \tag{8.65}$$

where $\hat{\mathbf{p}} = -i\hbar\nabla$, $\phi(\mathbf{r}, t)$ is the scalar potential and $\mathbf{A}(\mathbf{r}, t)$ the vector potential.
    To workout the non-relativistic limit of Eq. (8.65) it is useful to set

$$\Psi(\mathbf{r}, t) = e^{-imc^2 t/\hbar} \begin{pmatrix} \psi_1(\mathbf{r}, t) \\ \psi_2(\mathbf{r}, t) \\ \chi_1(\mathbf{r}, t) \\ \chi_2(\mathbf{r}, t) \end{pmatrix} = e^{-imc^2 t/\hbar} \begin{pmatrix} \psi(\mathbf{r}, t) \\ \chi(\mathbf{r}, t) \end{pmatrix} , \tag{8.66}$$

where $\psi(\mathbf{r}, t)$ and $\chi(\mathbf{r}, t)$ are two-component spinors, for which we obtain

$$i\hbar\frac{\partial}{\partial t}\begin{pmatrix}\psi \\ \chi\end{pmatrix} = \begin{pmatrix} q\,\phi & c\,\hat{\sigma}\cdot(\hat{\mathbf{p}}-q\mathbf{A}) \\ c\,\hat{\sigma}\cdot(\hat{\mathbf{p}}-q\mathbf{A}) & q\,\phi-2\,mc^2 \end{pmatrix}\begin{pmatrix}\psi \\ \chi\end{pmatrix} \tag{8.67}$$

where $\hat{\sigma} = (\hat{\sigma}_1, \hat{\sigma}_2, \hat{\sigma}_3)$. Remarkably, only in the lower equation of the system it appears the mass term mc^2, which is dominant in the non-relativistic limit. Indeed, under the approximation $\left(i\hbar\frac{\partial}{\partial t} - q\,\phi + 2mc^2\right)\chi \simeq 2mc^2\,\chi$, the previous equations become

$$\begin{pmatrix}i\hbar\frac{\partial\psi}{\partial t} \\ 0\end{pmatrix} = \begin{pmatrix} q\,\phi & c\,\hat{\sigma}\cdot(\hat{\mathbf{p}}-q\mathbf{A}) \\ c\,\hat{\sigma}\cdot(\hat{\mathbf{p}}-q\mathbf{A}) & -2\,mc^2 \end{pmatrix}\begin{pmatrix}\psi \\ \chi\end{pmatrix}, \tag{8.68}$$

from which

$$\chi = \frac{\hat{\sigma}\cdot(\hat{\mathbf{p}}-q\mathbf{A})}{2\,mc}\,\psi. \tag{8.69}$$

Inserting this expression in the upper equation of the system (8.68) we find

$$i\hbar\frac{\partial}{\partial t}\psi = \left(\frac{[\hat{\sigma}\cdot(\hat{\mathbf{p}}-q\mathbf{A})]^2}{2\,m} + q\,\phi\right)\psi. \tag{8.70}$$

From the identity

$$\left[\hat{\sigma}\cdot(\hat{\mathbf{p}}-q\mathbf{A})\right]^2 = (\hat{\mathbf{p}}-q\mathbf{A})^2 - i\,q\,(\hat{\mathbf{p}}\wedge\mathbf{A})\cdot\hat{\sigma} \tag{8.71}$$

where $\hat{\mathbf{p}} = -i\hbar\nabla$, and using the relation $\mathbf{B} = \nabla\wedge\mathbf{A}$ which introduces the magnetic field we finally get

$$i\hbar\frac{\partial}{\partial t}\psi(\mathbf{r}, t) = \left(\frac{(-i\hbar\nabla - q\mathbf{A}(\mathbf{r}, t))^2}{2\,m} - \frac{q}{m}\mathbf{B}(\mathbf{r}, t)\cdot\hat{\mathbf{S}} + q\,\phi(\mathbf{r}, t)\right)\psi(\mathbf{r}, t), \tag{8.72}$$

that is the so-called Pauli equation with

$$\hat{\mathbf{S}} = \frac{\hbar}{2}\hat{\sigma}. \tag{8.73}$$

the spin operator. This equation was introduced in 1927 (a year before the Dirac equation) by Wolfgang Pauli as an extension of the Schrödinger equation with the phenomenological inclusion of the spin operator. If the magnetic field $\mathbf{B}$ is constant, the vector potential can be written as

$$\mathbf{A} = \frac{1}{2}\mathbf{B}\wedge\mathbf{r} \tag{8.74}$$

and then

$$\left(\hat{\mathbf{p}} - q\mathbf{A}\right)^2 = \hat{\mathbf{p}}^2 - 2q\mathbf{A} \cdot \hat{\mathbf{p}} + q^2\mathbf{A}^2 = \hat{\mathbf{p}}^2 - q\mathbf{B} \cdot \hat{\mathbf{L}} + q^2(\mathbf{B} \wedge \mathbf{r})^2 , \qquad (8.75)$$

with $\hat{\mathbf{L}} = \mathbf{r} \wedge \hat{\mathbf{p}}$ the orbital angular momentum operator. Thus, the Pauli equation for a particle of charge q in a constant magnetic field reads

$$i\hbar\frac{\partial}{\partial t}\psi(\mathbf{r}, t) = \left(-\frac{\hbar^2\nabla^2}{2\,m} - \frac{q}{2m}\mathbf{B} \cdot \left(\hat{\mathbf{L}} + 2\hat{\mathbf{S}}\right) + \frac{q^2}{2m}(\mathbf{B} \wedge \mathbf{r})^2 + q\,\phi(\mathbf{r}, t)\right)\psi(\mathbf{r}, t).$$
$$(8.76)$$

In conclusion, we have shown that the spin $\hat{\mathbf{S}}$ naturally emerges from the Dirac equation. Moreover, the Dirac equation predicts very accurately the magnetic moment μ_S of the electron ($q = -e$, $m = m_e$) which appears in the spin energy $E_s = -\hat{\mu}_S \cdot \mathbf{B}$ where

$$\mu_S = -g_e\frac{\mu_B}{\hbar}\hat{\mathbf{S}} \qquad (8.77)$$

with gyromagnetic ratio $g_e = 2$ and Bohr magneton $\mu_B = e\hbar/(2m) \simeq 5.79 \cdot 10^{-5}$ eV/T.

8.3.2   Dirac Equation with a Central Potential

We now consider the stationary Dirac equation with a confining spherically-symmetric potential $V(r) = V(|\mathbf{r}|)$, namely

$$\left(-i\hbar c\,\hat{\boldsymbol{\alpha}} \cdot \nabla + \hat{\beta}\,mc^2 + V(r)\right)\Phi(\mathbf{r}) = E\,\Phi(\mathbf{r}). \qquad (8.78)$$

This equation is easily derived from Eq. (8.65) setting $\mathbf{A} = 0$, $q\phi = V(r)$, and

$$\Psi(\mathbf{r}, t) = e^{-iEt/\hbar}\,\Phi(\mathbf{r}). \qquad (8.79)$$

The relativistic Hamiltonian

$$\hat{H} = -i\hbar c\,\hat{\boldsymbol{\alpha}} \cdot \nabla + \hat{\beta}\,mc^2 + V(r) \qquad (8.80)$$

commutes with the total angular momentum operator

$$\hat{\mathbf{J}} = \hat{\mathbf{L}} + \hat{\mathbf{S}} = \mathbf{r} \wedge \hat{\mathbf{p}} + \frac{\hbar}{2}\hat{\boldsymbol{\sigma}} \qquad (8.81)$$

because the external potential is spherically symmetric. In fact, one can show that

$$[\hat{H}, \hat{\mathbf{L}}] = -i\hbar c\,\hat{\boldsymbol{\alpha}} \wedge \hat{\mathbf{p}} = -[\hat{H}, \hat{\mathbf{S}}]. \qquad (8.82)$$

Consequently one has

$$[\hat{H}, \hat{\mathbf{J}}] = \mathbf{0},\tag{8.83}$$

and also

$$[\hat{H}, \hat{J}^2] = 0, \qquad [\hat{J}^2, \hat{J}_x] = [\hat{J}^2, \hat{J}_y] = [\hat{J}^2, \hat{J}_z] = 0,\tag{8.84}$$

where the three components $\hat{J}_x, \hat{J}_y, \hat{J}_z$ of the total angular momentum $\hat{\mathbf{J}} = (\hat{J}_x, \hat{J}_y, \hat{J}_z)$ satisfy the familiar commutation relations

$$[\hat{J}_i, \hat{J}_j] = i\hbar\, \epsilon_{ijk}\, \hat{J}_k\tag{8.85}$$

with

$$\epsilon_{ijk} = \begin{cases} 1 & \text{if } (i, j, k) \text{ is } (x, y, z) \text{ or } (z, x, y) \text{ or } (y, z, x) \\ -1 & \text{if } (i, j, k) \text{ is } (x, z, y) \text{ or } (z, y, x) \text{ or } (y, x, z) \\ 0 & \text{if } i = j \text{ or } i = k \text{ or } j = k \end{cases}\tag{8.86}$$

the Levi-Civita symbol (also called Ricci-Curbastro symbol). Note that these commutation relations can be symbolically synthesized as

$$\hat{\mathbf{J}} \wedge \hat{\mathbf{J}} = i\hbar\, \hat{\mathbf{J}}.\tag{8.87}$$

Indicating the states which are simultaneous eigenstates of $\hat{H}$, $\hat{J}^2$ and $\hat{J}_z$ as $|njm_j\rangle$, one has

$$\hat{H}|njm_j\rangle = E_{nj}\,|njm_j\rangle,\tag{8.88}$$

$$\hat{J}^2|njm_j\rangle = \hbar^2 j(j+1)\,|njm_j\rangle,\tag{8.89}$$

$$\hat{J}_z|njm_j\rangle = \hbar m_j\,|njm_j\rangle,\tag{8.90}$$

where j is the quantum number of the total angular momentum and $m_j = -j, -j + 1, -j + 2, \ldots, j - 2, j - 1, j$ the quantum number of the third component of the total angular momentum.

In conclusion, we have found that the orbital angular momentum $\hat{\mathbf{L}}$ and the spin $\hat{\mathbf{S}}$ are not constants of motion of a particle in a central potential. On the contrary, the total angular momentum $\hat{\mathbf{J}} = \hat{\mathbf{L}} + \hat{\mathbf{S}}$ is a constant of motion, and also $\hat{J}^2$ and $\hat{J}_z$ are constants of motion.

8.3.3 Relativistic Hydrogen Atom and Fine Splitting

Let us consider now the electron of the hydrogen atom. We set $q = -e, m = m_e$ and

$$V(r) = -\frac{e^2}{4\pi\varepsilon_0|\mathbf{r}|} = -\frac{e^2}{4\pi\varepsilon_0 r}.\tag{8.91}$$

Then the eigenvalues E_{nj} of $\hat{H}$ are found to be

$$E_{nj} = \frac{mc^2}{\sqrt{1 + \frac{\alpha^2}{\left(n-j-\frac{1}{2}+\sqrt{(j+\frac{1}{2})^2-\alpha^2}\right)^2}}} - mc^2 , \qquad (8.92)$$

with $\alpha = e^2/(4\pi\epsilon_0\hbar c) \simeq 1/137$ the fine-structure constant. We do not prove this remarkable quantization formula, obtained independently in 1928 by Charles Galton Darwin and Walter Gordon, but we stress that expanding it in powers of the fine-structure constant α to order α^4 one gets

$$E_{nj} = E_n^{(0)}\left[1 + \frac{\alpha^2}{n}\left(\frac{1}{j+\frac{1}{2}} - \frac{3}{4n}\right)\right] , \qquad (8.93)$$

where

$$E_n^{(0)} = -\frac{1}{2}mc^2\frac{\alpha^2}{n^2} = -\frac{13.6\,\text{eV}}{n^2} \qquad (8.94)$$

is the familiar Bohr quantization formula of the non relativistic hydrogen atom. The term which corrects the Bohr formula, given by

$$\Delta E = E_n^{(0)}\frac{\alpha^2}{n}\left(\frac{1}{j+\frac{1}{2}} - \frac{3}{4n}\right) , \qquad (8.95)$$

is called fine splitting correction. This term removes the non relativistic degeneracy of energy levels, but not completely: double-degenerate levels remain with the same quantum numbers n and j but different orbital quantum number $l = j \pm 1/2$.

We have seen that, strictly speaking, in the relativistic hydrogen atom nor the orbital angular momentum $\hat{\mathbf{L}}$ nor the spin $\hat{\mathbf{S}}$ are constants of motion. As a consequence l, m_l, s and m_s are not good quantum numbers. Nevertheless, in practice, due to the smallness of fine-splitting corrections, one often assumes without problems that both $\hat{\mathbf{L}}$ and $\hat{\mathbf{S}}$ are approximately constants of motion.

As an example, let us calculate the fine splitting for the state $|3p\rangle$ of the hydrogen atom. The integer number j is the quantum number of the total angular momentum $\mathbf{J} = \mathbf{L} + \mathbf{S}$, where $j = 1/2$ if $l = 0$ and $j = l - 1/2, l + 1/2$ if $l \neq 0$. The state $|3p\rangle$ means $n = 3$ and $l = 1$, consequently $j = 1/2$ or $j = 3/2$. The hyperfine correction for $j = 1/2$ reads

$$\Delta E_{3,\frac{1}{2}} = E_3^{(0)}\frac{\alpha^2}{3}\left(1 - \frac{1}{4}\right) = E_3^{(0)}\frac{\alpha^2}{4} = -\frac{13.6\,\text{eV}}{9}\frac{1}{137^2\cdot 4} = -2.01\cdot 10^{-5}\,\text{eV} . \qquad (8.96)$$

The hyperfine correction for $j = 3/2$ is instead

$$\Delta E_{3,\frac{3}{2}} = E_3^{(0)} \frac{\alpha^2}{3} \left(\frac{1}{2} - \frac{1}{4} \right) = E_3^{(0)} \frac{\alpha^2}{12} = -\frac{13.6 \, \text{eV}}{9} \frac{1}{137^2 \cdot 12} = -6.71 \cdot 10^{-6} \, \text{eV} \,.$$

(8.97)

8.3.4 Relativistic Corrections to the Schrödinger Hamiltonian

It is important to stress that the relativistic Hamiltonian $\hat{H}$ of the Dirac equation in a spherically-symmetric potential, given by Eq. (8.80), can be expressed as the familiar non relativistic Schrödinger Hamiltonian

$$\hat{H}_0 = -\frac{\hbar^2}{2m} \nabla^2 + V(r)$$

(8.98)

plus an infinite sum of relativistic quantum corrections. To this aim one can start from the Dirac equation, written in terms of bi-spinors, i.e. Eq. (8.67) with $\mathbf{A}(\mathbf{r}, t) = \mathbf{0}$ and $q\phi(\mathbf{r}, t) = V(r)$, which gives

$$E \begin{pmatrix} \tilde{\psi} \\ \tilde{\chi} \end{pmatrix} = \begin{pmatrix} V(r) & c \, \hat{\boldsymbol{\sigma}} \cdot \hat{\mathbf{p}} \\ c \, \hat{\boldsymbol{\sigma}} \cdot \hat{\mathbf{p}} & V(r) - 2mc^2 \end{pmatrix} \begin{pmatrix} \tilde{\psi} \\ \tilde{\chi} \end{pmatrix}$$

(8.99)

setting $\psi(\mathbf{r}, t) = \tilde{\psi}(\mathbf{r}) \, e^{-iEt/\hbar}$ and $\chi(\mathbf{r}, t) = \tilde{\chi}(\mathbf{r}) \, e^{-iEt/\hbar}$. The lower equation of this system can be written as

$$\tilde{\chi} = \frac{c \, \hat{\boldsymbol{\sigma}} \cdot \hat{\mathbf{p}}}{E - V(r) + 2mc^2} \tilde{\psi} \,.$$

(8.100)

This is an exact equation. If $E - V(r) \ll 2mc^2$ the equation becomes

$$\tilde{\chi} = \frac{\hat{\boldsymbol{\sigma}} \cdot \hat{\mathbf{p}}}{2mc} \tilde{\psi} \,,$$

(8.101)

which is exactly the stationary version of Eq. (8.69) with $\mathbf{A}(\mathbf{r}, t) = \mathbf{0}$. We can do something better by expanding Eq. (8.100) with respect to the small term $(E - V(r))/2mc^2$ obtaining

$$\tilde{\chi} = \frac{\hat{\boldsymbol{\sigma}} \cdot \hat{\mathbf{p}}}{2mc} \left(1 - \frac{E - V(r)}{2mc^2} + \dots \right) \tilde{\psi} \,.$$

(8.102)

Inserting this expression in the upper equation of the system and neglecting the higher order terms symbolized by the three dots, and after some tedious calculations, one finds

$$E \tilde{\psi} = \hat{H} \tilde{\psi} \,,$$

(8.103)

where

$$\hat{H} = \hat{H}_0 + \hat{H}_1 + \hat{H}_2 + \hat{H}_3 , \qquad (8.104)$$

with

$$\hat{H}_1 = -\frac{\hbar^4}{8m^3c^2}\nabla^4 , \qquad (8.105)$$

$$\hat{H}_2 = \frac{1}{2m^2c^2}\frac{1}{r}\frac{dV(r)}{dr}\mathbf{L}\cdot\mathbf{S} , \qquad (8.106)$$

$$\hat{H}_3 = \frac{\hbar^2}{8m^2c^2}\nabla^2V(r) , \qquad (8.107)$$

with $\hat{H}_1$ the relativistic correction to the electron kinetic energy, $\hat{H}_2$ the spin-orbit correction, and $\hat{H}_3$ the Darwin correction.

If the external potential V(r) is that of the hydrogen atom, i.e. $V(r) = -e^2/(4\pi\varepsilon_0|\mathbf{r}|)$, one finds immediately that $H_3 = (\hbar^2e^2)/(8m^2c^2\varepsilon_0)\delta(\mathbf{r})$ because $\nabla^2(1/|\mathbf{r}|) = -4\pi\delta(\mathbf{r})$. In addition, by applying the first-order perturbation theory to $\hat{H}$ with $\hat{H}_0$ unperturbed Hamiltonian, one gets exactly Eq. (8.93) of fine-structure correction. Physically one can say that the relativistic fine structure is due to the coupling between the spin $\hat{\mathbf{S}}$ and the orbital angular momentum $\hat{\mathbf{L}}$ of the electron. Moreover, we observe that

$$\hat{\mathbf{L}}\cdot\hat{\mathbf{S}} = \frac{1}{2}\left(\hat{J}^2 - \hat{L}^2 - \hat{S}^2\right) , \qquad (8.108)$$

since

$$\hat{J}^2 = (\hat{\mathbf{L}} + \hat{\mathbf{S}})^2 = \hat{L}^2 + \hat{S}^2 + 2\,\hat{\mathbf{L}}\cdot\hat{\mathbf{S}} , \qquad (8.109)$$

the Hamiltonian $\hat{H}$ of Eq. (8.104) commutes with $\hat{L}^2$ and $\hat{S}^2$ but not with $\hat{L}_z$ and $\hat{S}_z$.

Actually, also the nucleus (the proton in the case of the hydrogen atom) has its spin $\hat{\mathbf{I}}$ which couples to the electronic spin to produce the so-called hyperfine structure. However, typically, hyperfine structure has energy shifts orders of magnitude smaller than the fine structure.

8.4   Spin Properties in a Magnetic Field

The amazing properties of quantum mechanics are very well illustrated by the dynamics of a spin in the presence of a magnetic field. As an example, let us consider the electron in a uniform magnetic field $\mathbf{B} = (0, 0, B_0)$. We want to calculate the expectation value of the spin $\hat{\mathbf{S}}$ along the x axis if at $t = 0$ the spin is along the z axis. The Hamiltonian operator of the spin is

$$\hat{H} = -\hat{\mu}_S \cdot \mathbf{B} , \qquad (8.110)$$

where $\hat{\mu}$ is the magnetic dipole moment of the electron, given by

$$\mu_S = -g_e \frac{\mu_B}{\hbar} \hat{S} = -\frac{1}{2} g_e \mu_B \hat{\sigma} , \qquad (8.111)$$

with $g_e = 2.002319 \simeq 2$ the gyromagnetic ratio of the electron, $\mu_B = e\hbar/(2m) = 9.27 \cdot 10^{-24} \, J/T$ the Bohr magneton and $\hat{\sigma} = (\hat{\sigma}_1, \hat{\sigma}_2, \hat{\sigma}_3)$ the vector of Pauli matrices. The Hamiltonian operator can be written as

$$\hat{H} = \frac{1}{2} \hbar \omega_0 \, \hat{\sigma}_3 = \frac{1}{2} \hbar \omega_0 \begin{pmatrix} 1 & 0 \\ 0 & -1 \end{pmatrix} , \qquad (8.112)$$

where $\omega_0 = g_e \mu_B B_0/\hbar$ is the Larmor frequency of the system. Suppose that the initial state of the system is

$$|\psi(0)\rangle = \begin{pmatrix} 1 \\ 0 \end{pmatrix} = |\uparrow\rangle , \qquad (8.113)$$

while

$$\begin{pmatrix} 0 \\ 1 \end{pmatrix} = |\downarrow\rangle \qquad (8.114)$$

is the other eigenstate of $\hat{\sigma}_3$. Then, the state at time t is then given by

$$|\psi(t)\rangle = e^{-i\hat{H}t/\hbar}|\psi(0)\rangle = e^{-i\omega_0\hat{\sigma}_3 t/2}|\uparrow\rangle = e^{-i\omega_0 t/2}|\uparrow\rangle , \qquad (8.115)$$

because

$$\hat{\sigma}_3|\uparrow\rangle = |\uparrow\rangle \qquad (8.116)$$

and

$$e^{-i\omega_0\hat{\sigma}_3 t/2}|\uparrow\rangle = e^{-i\omega_0 t/2}|\uparrow\rangle . \qquad (8.117)$$

The expectation value at time t of the spin component along the x axis is then

$$\langle \hat{S}_x(t)\rangle = \langle\psi(t)|\frac{\hbar}{2}\hat{\sigma}_1|\psi(t)\rangle = \frac{\hbar}{2}\langle\uparrow |e^{i\omega_0 t/2}\hat{\sigma}_1 e^{-i\omega_0 t/2}|\uparrow\rangle = \frac{\hbar}{2}\langle\uparrow |\hat{\sigma}_1|\uparrow\rangle . \qquad (8.118)$$

Observing that

$$\hat{\sigma}_1|\uparrow\rangle = |\downarrow\rangle , \qquad (8.119)$$

and also

$$\langle\uparrow | \downarrow\rangle = 0 , \qquad (8.120)$$

we conclude that

$$\langle\hat{S}_x(t)\rangle = 0 . \qquad (8.121)$$

This means that if initially the spin is in the direction of the magnetic field it remains in that direction forever: it is a stationary state with a time-dependence only in the phase. Instead, the components of spin which are orthogonal to the magnetic field have always zero expectation value.

Let us consider another interesting example: the electron is set in a uniform magnetic field $\mathbf{B} = (0, 0, B_0)$. We want to calculate the expectation value of the spin $\hat{\mathbf{S}}$ along the x axis if at $t = 0$ the spin is along the x axis. The Hamiltonian operator of spin is given by

$$\hat{H} = \frac{1}{2}\hbar\omega_0\,\hat{\sigma}_3 \tag{8.122}$$

where $\omega_0 = g_e\mu_B B_0/\hbar$ is the Larmor frequency of the system, with g_e gyromagnetic factor and μ_B Bohr magneton. The initial state of the system is

$$|\psi(0)\rangle = \frac{1}{\sqrt{2}}\left(|\uparrow\rangle + |\downarrow\rangle\right)\,, \tag{8.123}$$

since

$$\hat{\sigma}_1|\psi(0)\rangle = \frac{1}{\sqrt{2}}\left(\hat{\sigma}_1|\uparrow\rangle + \sigma_1|\downarrow\rangle\right) = \frac{1}{\sqrt{2}}\left(|\downarrow\rangle + |\uparrow\rangle\right) = |\psi(0)\rangle\,. \tag{8.124}$$

The state at time t is then

$$|\psi(t)\rangle = e^{-i\hat{H}t/\hbar}|\psi(0)\rangle = e^{-i\omega_0\hat{\sigma}_3 t/2}\frac{1}{\sqrt{2}}\left(|\uparrow\rangle + |\downarrow\rangle\right) = \frac{1}{\sqrt{2}}\left(e^{-i\omega_0 t/2}|\uparrow\rangle + e^{i\omega_0 t/2}|\downarrow\rangle\right)\,, \tag{8.125}$$

because

$$\hat{\sigma}_3|\uparrow\rangle = |\uparrow\rangle \tag{8.126}$$

$$\hat{\sigma}_3|\downarrow\rangle = -|\downarrow\rangle \tag{8.127}$$

and

$$e^{-i\omega_0\hat{\sigma}_3 t/2}|\uparrow\rangle = e^{-i\omega_0 t/2}|\uparrow\rangle \tag{8.128}$$

$$e^{-i\omega_0\hat{\sigma}_3 t/2}|\downarrow\rangle = e^{i\omega_0 t/2}|\downarrow\rangle\,. \tag{8.129}$$

The expectation value at time t of the spin component along x axis reads

$$\langle\hat{S}_x(t)\rangle = \langle\psi(t)|\frac{\hbar}{2}\hat{\sigma}_1|\psi(t)\rangle = \frac{\hbar}{4}\left(\langle\uparrow|e^{i\omega_0 t/2} + \langle\downarrow|e^{-i\omega_0 t/2}\right)\hat{\sigma}_1\left(e^{-i\omega_0 t/2}|\uparrow\rangle + e^{i\omega_0 t/2}|\downarrow\rangle\right) \tag{8.130}$$

$$= \frac{\hbar}{4}\left(e^{i\omega_0 t}\langle\uparrow|\uparrow\rangle + e^{-i\omega_0 t}\langle\downarrow|\downarrow\rangle\right) = \frac{\hbar}{2}\cos\left(\omega_0 t\right)\,. \tag{8.131}$$

8.5 Stark Effect

Let us consider the hydrogen atom under the action of a constant electric field $\mathbf{E}$. We write the constant electric field as

$$\mathbf{E} = E\,\mathbf{u}_z = (0, 0, E)\,, \tag{8.132}$$

choosing the z axis in the same direction of $\mathbf{E}$. The Hamiltonian operator of the system is then given by

$$\hat{H} = \hat{H}_0 + \hat{H}_I\,, \tag{8.133}$$

where

$$\hat{H}_0 = \frac{\hat{p}^2}{2m} - \frac{e^2}{4\pi\epsilon_0\,r} \tag{8.134}$$

is the non-relativistic Hamiltonian of the electron in the hydrogen atom (with $m = m_e$ the electron mass), while

$$\hat{H}_I = -e\,\phi(\mathbf{r}) = e\,\mathbf{E}\cdot\mathbf{r} = -\mathbf{d}\cdot\mathbf{E} = e\,E\,z \tag{8.135}$$

is the Hamiltonian of the interaction due to the electric scalar potential $\phi(\mathbf{r}) = -\mathbf{E}\cdot\mathbf{r}$ such that $\mathbf{E} = -\nabla\phi$, with $-e$ electric charge of the electron and $\mathbf{d} = -e\mathbf{r}$ the electric dipole.

Let $|nlm_l\rangle$ be the eigenstates of $\hat{H}_0$, such that

$$\hat{H}_0|nlm_l\rangle = E_n^{(0)}|nlm_l\rangle \tag{8.136}$$

with

$$E_n^{(0)} = -\frac{13.6}{n^2}\,\text{eV}\,, \tag{8.137}$$

the Bohr spectrum of the hydrogen atom, and moreover

$$\hat{L}^2|nlm_l\rangle = \hbar^2\,l(l+1)\,|nlm_l\rangle\,, \qquad \hat{L}_z|nlm_l\rangle = \hbar\,m_l\,|nlm_l\rangle\,. \tag{8.138}$$

At the first order of degenerate perturbation theory the energy spectrum is given by

$$E_n = E_n^{(0)} + E_n^{(1)}\,, \tag{8.139}$$

where $E_n^{(1)}$ is one of the eigenvalues of the submatrix $\hat{H}_I^n$, whose elements are

$$a^{(n)}_{l'm_l',lm_l} = \langle nl'm_l'|\hat{H}_I|nlm_l\rangle = e\,E\,\langle nl'm_l'|z|nlm_l\rangle\,. \tag{8.140}$$

Thus, in general, there is a linear splitting of degenerate energy levels due to the external electric field E. This effect is named after Johannes Stark, who discovered it in 1913. Actually, it was discovered independently in the same year also by Antonino Lo Surdo.

It is important to stress that the ground-state $|1s\rangle = |n = 1, l = 0, m_l = 0\rangle$ of the hydrogen atom is not degenerate and for it $\langle 1s|z|1s\rangle = 0$. It follows that there is no linear Stark effect for the ground-state. Thus, we need the second order perturbation theory, namely

$$E_1 = E_1^{(0)} + E_1^{(1)} + E_1^{(2)} ,\tag{8.141}$$

where

$$E_1^{(1)} = \langle 100|eEz|100\rangle = 0 ,\tag{8.142}$$

$$E_1^{(2)} = \sum_{nlm_l \neq 100} \frac{|\langle nlm_l|eEz|100\rangle|^2}{E_1^{(0)} - E_n^{(0)}} = e^2 E^2 \sum_{n=2}^{\infty} \frac{|\langle n10|z|100\rangle|^2}{E_1^{(0)} - E_n^{(0)}} ,\tag{8.143}$$

where the last equality is due to the dipole selection rules. This formula shows that the electric field produces a quadratic shift in the energy of the ground state. This phenomenon is known as the quadratic Stark effect.

The polarizability α_p of an atom is defined in terms of the energy-shift ΔE_1 of the atomic ground state energy E_1 induced by an external electric field E as follows:

$$\Delta E_1 = -\frac{1}{2} \alpha_p E^2 .\tag{8.144}$$

Hence, for the hydrogen atom we can write

$$\alpha_p = -2e^2 \sum_{n=2}^{\infty} \frac{|\langle n10|z|100\rangle|^2}{E_1^{(0)} - E_n^{(0)}} = -\frac{9}{4} \frac{e^2 r_0^2}{E_1^{(0)}} ,\tag{8.145}$$

with r_0 the Bohr radius.

8.6 Zeeman Effect

Let us consider the hydrogen atom under the action of a constant magnetic field $\mathbf{B}$. According to the Pauli equation, the Hamiltonian operator of the system is given by

$$\hat{H} = \frac{(\hat{\mathbf{p}} + e\mathbf{A})^2}{2m} - \frac{e^2}{4\pi\epsilon_0 r} - \hat{\boldsymbol{\mu}}_S \cdot \mathbf{B}\tag{8.146}$$

where

$$\hat{\boldsymbol{\mu}}_S = -\frac{e}{m}\hat{\mathbf{S}}\tag{8.147}$$

is the spin dipole magnetic moment, with $\hat{\mathbf{S}}$ the spin of the electron, and $\mathbf{A}$ is the vector potential, such that $\mathbf{B} = \nabla \wedge \mathbf{A}$. Because the magnetic field $\mathbf{B}$ is constant, the vector potential can be written as

$$\mathbf{A} = \frac{1}{2}\mathbf{B} \wedge \mathbf{r} , \tag{8.148}$$

and then

$$\left(\hat{\mathbf{p}} + e\mathbf{A}\right)^2 = \hat{\mathbf{p}}^2 + 2e\hat{\mathbf{p}} \cdot \mathbf{A} + e^2\mathbf{A}^2 = \hat{\mathbf{p}}^2 + 2e\mathbf{B} \cdot \hat{\mathbf{L}} + e^2(\mathbf{B} \wedge \mathbf{r})^2 , \tag{8.149}$$

with $\hat{\mathbf{L}} = \mathbf{r} \wedge \hat{\mathbf{p}}$. In this way the Hamiltonian can be expressed as

$$\hat{H} = \hat{H}_0 + \hat{H}_I , \tag{8.150}$$

where

$$\hat{H}_0 = \frac{\hat{p}^2}{2m} - \frac{e^2}{4\pi\epsilon_0 r} \tag{8.151}$$

is the non-relativistic Hamiltonian of the electron in the hydrogen atom, while

$$\hat{H}_I = -\hat{\boldsymbol{\mu}} \cdot \mathbf{B} + \frac{e^2}{8m}(\mathbf{B} \wedge \mathbf{r})^2 \tag{8.152}$$

is the Hamiltonian of the magnetic interaction, with

$$\hat{\boldsymbol{\mu}} = \hat{\boldsymbol{\mu}}_L + \hat{\boldsymbol{\mu}}_S = -\frac{e}{2m}(\hat{\mathbf{L}} + 2\hat{\mathbf{S}}) \tag{8.153}$$

the total dipole magnetic moment of the electron. We now write the constant magnetic field as

$$\mathbf{B} = B \, \mathbf{u}_z = (0, 0, B) , \tag{8.154}$$

choosing the z axis in the same direction of $\mathbf{B}$. In this way the interaction Hamiltonian becomes

$$\hat{H}_I = \frac{eB}{2m}\left(\hat{L}_z + 2\hat{S}_z\right) + \frac{e^2 B^2}{8m}(x^2 + y^2) . \tag{8.155}$$

The first term, called paramagnetic term, grows linearly with the magnetic field B while the second one, the diamagnetic term, grows quadratically. The paramagnetic term is of the order of $\mu_B B$, where $\mu_B = e\hbar/(2m) = 9.3 \cdot 10^{-24}$ Joule/Tesla $= 5.29 \cdot 10^{-5}$ eV/Tesla is the Bohr magneton. Because the unperturbed energy of $\hat{H}_0$ is of the order of 10 eV, the paramagnetic term can be considered a small perturbation.

8.6.1   Strong-Field Zeeman Effect

Usually the diamagnetic term is much smaller than the paramagnetic one, and becomes observable only for B of the order of $10^6/n^4$ Tesla, i.e. mainly in the astrophysical context. Thus in laboratory the diamagnetic term is usually negligible (apart for very large values of the principal quantum number n) and the effective interaction Hamiltonian reads

$$H_I = \frac{eB}{2m} \left(\hat{L}_z + 2\hat{S}_z \right) . \tag{8.156}$$

Thus (8.151) is the unperturbed Hamiltonian and (8.156) the perturbing Hamiltonian. It is clear that this total Hamiltonian is diagonal with respect to the eigenstates $|nlm_l m_s\rangle$ and one obtains immediately the following energy spectrum

$$E_{n,m_l,m_s} = E_n^{(0)} + \mu_B B \left(m_l + 2m_s \right) , \tag{8.157}$$

where $E_n^{(0)}$ is the unperturbed Bohr eigenspectrum and $\mu_B = e\hbar/(2m)$ is the Bohr magneton, with m the mass of the electron. Equation (8.157) describes the high-field Zeeman effect, first observed in 1896 by Pieter Zeeman. The field B does not remove the degeneracy in l but it does remove the degeneracy in m_l and m_s. The selection rules for dipolar transitions require $\Delta m_s = 0$ and $\Delta m_l = 0, \pm 1$. Thus the spectral line corresponding to a transition $n \to n'$ is split into 3 components, called Lorentz triplet.

8.6.2   Weak-Field Zeeman Effect

In the hydrogen atom the strong-field Zeeman effect is observable if the magnetic field B is between about $1/n^3$ Tesla and about $10^6/n^4$ Tesla, with n the principal quantum number. In fact, as previously explained, for B larger than about $10^6/n^4$ Tesla the diamagnetic term is no more negligible. Instead, for B smaller than about $1/n^3$ Tesla the splitting due to the magnetic field B becomes comparable with the splitting due to relativistic fine-structure corrections. Thus, to study the effect of a weak field B, i.e. the weak-field Zeeman effect, the unperturbed non-relativistic Hamiltonian $\hat{H}_0$ given by the Eq. (8.151) is no more reliable. One must use instead the exact relativistic Hamiltonian or, at least, the non-relativistic one with relativistic corrections, namely

$$\hat{H}_0 = \hat{H}_{0,0} + \hat{H}_{0,1} + \hat{H}_{0,2} + \hat{H}_{0,3} , \tag{8.158}$$

where

$$\hat{H}_{0,0} = -\frac{\hbar^2}{2m}\nabla^2 - \frac{e^2}{4\pi\epsilon_0 r} \tag{8.159}$$

$$\hat{H}_{0,1} = -\frac{\hbar^4}{8m^3c^2}\nabla^4 , \tag{8.160}$$

$$\hat{H}_{0,2} = \frac{1}{2m^2c^2}\frac{1}{r}\frac{dV(r)}{dr}\mathbf{L}\cdot\mathbf{S} , \tag{8.161}$$

$$\hat{H}_{0,3} = \frac{\hbar^2}{8m^2c^2}\nabla^2 V(r) , \tag{8.162}$$

with $\hat{H}_{0,0}$ the non relativistic Hamiltonian, $\hat{H}_{0,1}$ the relativistic correction to the electron kinetic energy, $\hat{H}_{0,2}$ the spin-orbit correction, and $\hat{H}_{0,3}$ the Darwin correction. In any case, m_l and m_s are no more good quantum numbers because $\hat{L}_z$ and $\hat{S}_z$ do not commute with the new $\hat{H}_0$.

The Hamiltonian (8.158) commutes instead with $\hat{L}^2$, $\hat{S}^2$, $\hat{J}^2$ and $\hat{J}_z$. Consequently, for this Hamiltonian the good quantum numbers are n, l, s, j and m_j. Applying again the first-order perturbation theory, where now (8.158) is the unperturbed Hamiltonian and (8.156) is the perturbing Hamiltonian, one obtains the following energy spectrum

$$E_{n,l,j,m_j} = E_{n,j}^{(0)} + E_{n,l,s,j,m_j}^{(1)} , \tag{8.163}$$

where $E_{n,j}^{(0)}$ is the unperturbed relativistic spectrum and

$$E_{n,l,s,j,m_j}^{(1)} = \frac{eB}{2m}\langle n, l, s, j, m_j|\hat{L}_z + 2\hat{S}_z|n, l, s, j, m_j\rangle \tag{8.164}$$

is the first-order correction, which is indeed not very easy to calculate. But we can do it. First we note that

$$\langle n, l, s, j, m_j|\hat{L}_z + 2\hat{S}_z|n, l, s, j, m_j\rangle = \langle n, l, s, j, m_j|\hat{J}_z + \hat{S}_z|n, l, s, j, m_j\rangle$$
$$= \hbar m_j + \langle n, l, s, j, m_j|\hat{S}_z|n, l, s, j, m_j\rangle . \tag{8.165}$$

The Wigner-Eckart theorem states that for any vector operator $\hat{\mathbf{V}} = (\hat{V}_1, \hat{V}_2, \hat{V}_3)$ such that $[\hat{J}_i, \hat{V}_j] = i\hbar\epsilon_{ijk}\hat{V}_k$ holds the identity

$$\hbar^2 j(j+1)\langle n, l, s, j, m_j|\hat{\mathbf{V}}|n, l, s, j, m_j\rangle = \langle n, l, s, j, m_j|(\hat{\mathbf{V}}\cdot\hat{\mathbf{J}})\hat{\mathbf{J}}|n, l, s, j, m_j\rangle . \tag{8.166}$$

In our case $\hat{\mathbf{V}} = \hat{\mathbf{S}}$ and we have considered the z component only. Then, on the basis of the Wigner-Eckart theorem, we have

$$\hbar^2 j(j+1)\langle n, l, s, j, m_j|\hat{S}_z|n, l, s, j, m_j\rangle = \langle n, l, s, j, m_j|(\hat{\mathbf{S}}\cdot\hat{\mathbf{J}})\hat{J}_z|n, l, s, j, m_j\rangle , \tag{8.167}$$

from which we obtain

$$\hbar^2 j(j+1)\langle n,l,s,j,m_j|\hat{S}_z|n,l,s,j,m_j\rangle = \hbar m_j\langle n,l,s,j,m_j|\hat{\mathbf{S}}\cdot\hat{\mathbf{J}}|n,l,s,j,m_j\rangle$$

$$= \hbar m_j\langle n,l,s,j,m_j|\frac{1}{2}\left(\hat{J}^2+\hat{S}^2-\hat{L}^2\right)|n,l,s,j,m_j\rangle$$

$$= \hbar m_j\frac{1}{2}\hbar^2\left(j(j+1)+s(s+1)-l(l+1)\right). \quad (8.168)$$

In conclusion, for a weak magnetic field B the first-order correction is given by

$$E^{(1)}_{n,l,s,j,m_j} = \mu_B\, B\, g_{l,s,j}\, m_j\,, \qquad (8.169)$$

where $\mu_B = e\hbar/(2m)$ is the Bohr magneton, and

$$g_{l,s,j} = 1 + \frac{j(j+1)+s(s+1)-l(l+1)}{2j(j+1)} \qquad (8.170)$$

is the so-called Landé factor. Clearly, in the case of the electron $s = 1/2$ and the Landé factor becomes

$$g_{l,j} = 1 + \frac{j(j+1)-l(l+1)+3/4}{2j(j+1)}\,. \qquad (8.171)$$

Strictly speaking, the energy splitting described by Eq. (8.169) is fully reliable only for a weak magnetic field B in the range 0 Tesla $\leq B \ll 1/n^3$ Tesla, with n the principal quantum number. In fact, if the magnetic field B approaches $1/n^3$ Tesla one observes a complex pattern of splitting, which moves by increasing B from the splitting described by Eq. (8.169) towards the splitting described by Eq. (8.157). This transition, observed in 1913 by Friedrich Paschen and Ernst Back, is now called the Paschen-Back effect.

Further Reading

A complete treatment of the Schrödinger equation for the electron in the hydrogen atom can be found in:
Landau, L.D., Lifshitz, E.M.: Quantum Mechanics: Non-Relativistic Theory. Pergamon (1981)
A detailed analysis of the selection rules based on the dipole approximation of the matter-radiation interaction can be found in:
Salasnich, L.: Quantum Physics of Light and Matter: Photons, Atoms, and Strongly Correlated Systems. Springer (2017)
For the Dirac equation and the Pauli equation:
Bjorken, J.D., Drell, S.D.: Relativistic Quantum Mechanics. McGraw-Hill, New York (1964)
A detailed discussion of the Dirac equation for the electron in the hydrogen atom can be found in:

Lifshitz, E.M., Pitaevskii, L.P.: Relativistic Quantum Theory. Pergamon (1971)
For the Stark effect:
Robinett, R.W.: Quantum Mechanics: Classical Results, Modern Systems, and Visualized Examples, chap. 19, Sect. 19.6. Oxford Univ. Press, Oxford (2006)
For the Zeeman effect:
Bransden, B.H., Joachain, C.J.: Physics of Atoms and Molecules, chap. 6, Sects. 6.1 and 6.2. Prentice Hall, Upper Saddle River (2003)

Chapter 9
Quantum Mechanics of Many-Body Systems

In this chapter we analyze atoms with many electrons and, more generally, systems with many interacting identical particles. We consider the general properties of many identical particles with their bosonic or fermionic many-body wavefunctions, and the connection between spin and statistics which explains the Pauli principle and the main features of the periodic table of elements.

9.1 Identical Quantum Particles

First of all, we introduce the generalized coordinate $x = (\mathbf{r}, \sigma)$ of a particle which takes into account the spatial coordinate $\mathbf{r}$ but also the intrinsic spin σ pertaining to the particle. For instance, a spin $1/2$ particle has $\sigma = -1/2, 1/2 = \downarrow, \uparrow$. By using the Dirac notation the corresponding single-particle state is

$$|x\rangle = |\mathbf{r}\,\sigma\rangle. \tag{9.1}$$

We now consider N identical particles; for instance particles with the same mass and electric charge. The many-body wavefunction of the system is given by

$$\Psi(x_1, x_2, \ldots, x_N) = \Psi(\mathbf{r}_1, \sigma_1, \mathbf{r}_2, \sigma_2, \ldots, \mathbf{r}_N, \sigma_N), \tag{9.2}$$

According to quantum mechanics identical particles are indistinguishable. As a consequence, it must be

$$|\Psi(x_1, x_2, \ldots, x_i, \ldots, x_j, \ldots, x_N)|^2 = |\Psi(x_1, x_2, \ldots, x_j, \ldots, x_i, \ldots, x_N)|^2, \tag{9.3}$$

© The Author(s), under exclusive license to Springer Nature Switzerland AG 2022
L. Salasnich, *Modern Physics*, UNITEXT for Physics,
https://doi.org/10.1007/978-3-030-93743-0_9

which means that the probability of finding the particles must be independent on the exchange of two generalized coordinates x_i and x_j. Obviously, for 2 particles this implies that

$$|\Psi(x_1, x_2)|^2 = |\Psi(x_2, x_1)|^2 \,. \tag{9.4}$$

Experiments suggests that there are only two kind of identical particles which satisfy Eq. (9.3): bosons and fermions. For N identical bosons one has

$$\Psi(x_1, x_2, \ldots, x_i, \ldots, x_j, \ldots, x_N) = \Psi(x_1, x_2, \ldots, x_j, \ldots, x_i, \ldots, x_N) \,, \tag{9.5}$$

i.e. the many-body wavefunction is symmetric with respect to the exchange of two coordinates x_i and x_j. Note that for 2 identical bosonic particles this implies

$$\Psi(x_1, x_2) = \Psi(x_2, x_1). \tag{9.6}$$

For N identical fermions one has instead

$$\Psi(x_1, x_2, \ldots, x_i, \ldots, x_j, \ldots, x_N) = -\Psi(x_1, x_2, \ldots, x_j, \ldots, x_i, \ldots, x_N) \,, \tag{9.7}$$

i.e. the many-body wavefunction is anti-symmetric with respect to the exchange of two coordinates x_i and x_j. Note that for 2 identical fermionic particles this implies

$$\Psi(x_1, x_2) = -\Psi(x_2, x_1) \,. \tag{9.8}$$

An immediate consequence of the anti-symmetry of the fermionic many-body wave-function is the *Pauli Principle*: if $x_i = x_j$ then the many-body wavefunction is zero. In other words: the probability of finding two fermionic particles with the same generalized coordinates is zero.

9.1.1   Spin-Statistics Theorem

A remarkable experimental fact, which is often called *spin-statistics theorem* because can be deduced from other postulates of relativistic quantum field theory, is the following: identical particles with integer spin are bosons while identical particles with semi-integer spin are fermions. For instance, photons are bosons with spin 1 while electrons are fermions with spin $1/2$. Notice that for a composed particle it is the total spin which determines the statistics. For example, the total spin (sum of nuclear and electronic spins) of ^{4}He atom is 0 and consequently this atom is a boson, while the total spin of ^{3}He atom is $1/2$ and consequently this atom is a fermion.

9.2 Non-interacting Identical Particles

The quantum Hamiltonian of N identical non-interacting particles is given by

$$\hat{H}_0 = \sum_{i=1}^{N} \hat{h}(x_i) \,, \tag{9.9}$$

where $\hat{h}(x)$ is the single-particle Hamiltonian. Usually the single-particle Hamiltonian is given by

$$\hat{h}(x) - -\frac{\hbar^2}{2m}\nabla^2 + U(\mathbf{r}) \,, \tag{9.10}$$

with $U(\mathbf{r})$ the external confining potential. In general the single-particle Hamiltonian $\hat{h}$ satisfies the eigenvalue equation

$$\hat{h}(x)\,\phi_n(x) = \epsilon_n\,\phi_n(x) \,, \tag{9.11}$$

where ϵ_n are the single-particle eigenenergies and $\phi_n(x)$ the single-particle eigenfunctions, with $n = 1, 2, \ldots$.

The many-body wavefunction $\Psi(x_1, x_2, \ldots, x_N)$ of the system can be written in terms of the single-particle wavefunctions $\phi_n(x)$ but one must take into account the spin-statistics of the identical particles. For N bosons the simplest many-body wave function reads

$$\Psi(x_1, x_2, \ldots, x_N) = \phi_1(x_1)\,\phi_1(x_2)\,\ldots\,\phi_1(x_N) \,, \tag{9.12}$$

which corresponds to the configuration where all the particles are in the lowest-energy single-particle state $\phi_1(x)$. This is indeed a pure Bose-Einstein condensate. Note that for 2 bosons the previous expression becomes

$$\Psi(x_1, x_2) = \phi_1(x_1)\phi_1(x_2) \,. \tag{9.13}$$

Obviously there are infinite configuration which satisfy the bosonic symmetry of the many-body wavefunction. For example, with 2 bosons one can have

$$\Psi(x_1, x_2) = \phi_4(x_1)\phi_4(x_2) \,, \tag{9.14}$$

which means that the two bosons are both in the fourth eigenstate; another example is

$$\Psi(x_1, x_2) = \frac{1}{\sqrt{2}}\left(\phi_1(x_1)\phi_2(x_2) + \phi_1(x_2)\phi_2(x_1)\right) \,, \tag{9.15}$$

where the factor $1/\sqrt{2}$ has been included to maintain the same normalization of the many-body wavefunction, and in this case the bosons are in the first two available single-particles eigenstates.

For N fermions the simplest many-body wave function is instead very different, and it is given by

$$\Psi(x_1, x_2, \ldots, x_N) = \frac{1}{\sqrt{N!}} \begin{pmatrix} \phi_1(x_1) & \phi_1(x_2) & \ldots & \phi_1(x_N) \\ \phi_2(x_1) & \phi_2(x_2) & \ldots & \phi_2(x_N) \\ \ldots & \ldots & \ldots & \ldots \\ \phi_N(x_1) & \phi_N(x_2) & \ldots & \phi_N(x_N) \end{pmatrix} \qquad (9.16)$$

that is the so-called Slater determinant of the $N \times N$ matrix obtained with the N lowest-energy single particle wavefunctions $\psi_n(x)$, with $n = 1, 2, \ldots, N$, calculated in the N possible generalized coordinates x_i, with $i = 1, 2, \ldots, N$. Note that for 2 fermions the previous expression becomes

$$\Psi(x_1, x_2) = \frac{1}{\sqrt{2}} \left(\phi_1(x_1)\phi_2(x_2) - \phi_1(x_2)\phi_2(x_1) \right) . \qquad (9.17)$$

We stress that for non-interacting identical particles the Hamiltonian (9.9) is separable and the total energy associated to the bosonic many-body wavefunction (9.12) is simply

$$E = N \epsilon_1 , \qquad (9.18)$$

while for the fermionic many-body wavefunction (9.16) the total energy (in the absence of degenerate single-particle energy levels and for spin-polarized fermions) reads

$$E = \epsilon_1 + \epsilon_2 + \cdots + \epsilon_N , \qquad (9.19)$$

which is surely higher than the bosonic one. The highest occupied single-particle energy level is called Fermi energy, and it indicated as ϵ_F; in our case it is obviously $\epsilon_F = \epsilon_N$.

9.2.1   Atomic Shell Structure and the Periodic Table of the Elements

The non-relativistic quantum Hamiltonian of Z identical non-interacting electrons in the neutral atom is given by

$$\hat{H}_0 = \sum_{i=1}^{Z} \hat{h}(\mathbf{r}_i) , \qquad (9.20)$$

where $\hat{h}(\mathbf{r})$ is the single-particle Hamiltonian given by

$$\hat{h}(\mathbf{r}) = -\frac{\hbar^2}{2m}\nabla^2 + U(\mathbf{r}), \qquad (9.21)$$

with

$$U(\mathbf{r}) = -\frac{Ze^2}{4\pi\varepsilon_0\,|\mathbf{r}|} \qquad (9.22)$$

the confining potential due to the attractive Coulomb interaction between the single electron and the atomic nucleus of positive charge Ze, with $e > 0$.

Because the confining potential $U(\mathbf{r})$ is spherically symmetric, i.e. $U(\mathbf{r}) = U(|\mathbf{r}|)$, the single-particle Hamiltonian $\hat{h}$ satisfies the eigenvalue equation

$$\hat{h}(\mathbf{r})\,\phi_{nlm_lm_s}(\mathbf{r}) = \epsilon_n(Z)\,\phi_{nlm_lm_s}(\mathbf{r}), \qquad (9.23)$$

where

$$\epsilon_n(Z) = -13.6\,eV\,\frac{Z^2}{n^2} \qquad (9.24)$$

are the Bohr single-particle eigenenergies of the hydrogen-like atom, and $\phi_{nlm_lm_s}(\mathbf{r})$ $= R_{nl}(r)Y_{lm_l}(\theta, \phi)$ are the single-particle eigenfunctions, which depends on the principal quantum numbers $n = 1, 2, \ldots$, the angular quantum number $l = 0, 1, \ldots, n - 1$, the third-component angular quantum number $m_l = -l, -l+1, \ldots, l-1, l$, and the third-component spin quantum number $m_s = -\frac{1}{2}, \frac{1}{2}$. Notice that here

$$\phi_{nlm_lm_s}(\mathbf{r}) = \phi_{nlm_l}(\mathbf{r}, \sigma) \qquad (9.25)$$

with $\sigma =\uparrow$ for $m_s = \frac{1}{2}$ and $\sigma =\downarrow$ for $m_s = -\frac{1}{2}$.

Z	Atom	Symbol	E
1	hydrogen	H	$\epsilon_1(1)$
2	helium	He	$2\epsilon_1(2)$
3	lithium	Li	$2\epsilon_1(3) + \epsilon_2(3)$
4	beryllium	Be	$2\epsilon_1(4) + 2\epsilon_2(4)$
5	boron	B	$2\epsilon_1(5) + 3\epsilon_2(5)$
6	carbon	C	$2\epsilon_1(6) + 4\epsilon_2(6)$
7	nitrogen	N	$2\epsilon_1(7) + 5\epsilon_2(7)$
8	oxygen	O	$2\epsilon_1(8) + 6\epsilon_2(8)$
9	fluorine	F	$2\epsilon_1(9) + 7\epsilon_2(9)$
10	neon	Ne	$2\epsilon_1(10) + 8\epsilon_2(10)$
11	sodium	Na	$2\epsilon_1(11) + 8\epsilon_2(11) + \epsilon_3(11)$
12	magnesium	Mg	$2\epsilon_1(12) + 8\epsilon_2(12) + 2\epsilon_3(12)$
13	aluminium	Al	$2\epsilon_1(13) + 8\epsilon_2(13) + 3\epsilon_3(13)$
14	silicon	Si	$2\epsilon_1(14) + 8\epsilon_2(14) + 4\epsilon_3(14)$

Table. Lightest atoms and their ground-state energy E on the basis of single-particle energies.

Due to the Pauli principle the ground-state energy E of this system of Z electrons strongly depends of the degeneracy of single-particle energy levels. In the Table we report the ground-state energy E of the lightest atoms on the basis of their single-particle energy levels $\epsilon_n(Z)$.

The degeneracy of the single-particle energy level $\epsilon_n(Z)$ is clearly independent on Z and given by

$$deg(\epsilon_n(Z)) = \sum_{l=0}^{n-1} 2(2l+1) = 2n^2, \qquad (9.26)$$

which is the maximum number of electrons with the same principal quantum number n. The set of states with the same principal quantum number is called theoretical shell. The number of electrons in each theoretical shell are: 2, 8, 18, 32, 52. One expects that the more stable atoms are characterized by fully occupied theoretical shells. Actually, the experimental data, namely the periodic table of elements due to Dmitri Mendeleev, suggest that the true number of electrons in each experimental shell are instead: 2, 8, 8, 18, 18, 32, because the noble atoms are characterized the following atomic numbers:

$$2, \ 2+8 = 10, \ 2+8+8 = 18, \ 2+8+8+18 = 36, \ 2+8+8+18+18 = 54, \ 2+8+8+18+18+32 = 86,$$

corresponding to Helium ($Z = 2$), Neon ($Z = 10$), Argon ($Z = 18$), Krypton ($Z = 36$), Xenon ($Z = 54$), and Radon ($Z = 86$). The experimental sequence is clearly similar but not equal to the theoretical one, due to repetitions of 8 and 18.

It is important to stress that the theoretical sequence is obtained under the very crude assumption of non-interacting electrons. To improve the agreement between theory and experiment one must include the interaction between the electrons.

9.3   Interacting Identical Particles

The quantum Hamiltonian of N identical interacting particles is given by

$$\hat{H} = \sum_{i=1}^{N} \hat{h}(x_i) + \frac{1}{2} \sum_{\substack{i,j=1 \\ i \neq j}}^{N} V(x_i, x_j) = \hat{H}_0 + \hat{H}_I, \qquad (9.27)$$

where $\hat{h}$ is the single-particle Hamiltonian and $V(x_i, x_j)$ is the inter-particle potential of the mutual interaction. In general, due to the inter-particle potential, the Hamiltonian (9.27) is not separable and the many-body wavefunctions given by Eqs. (9.12) and (9.16) are not exact eigenfunctions of $\hat{H}$.

9.3.1 Electrons in Atoms and Molecules

It is well known that a generic molecule is made of N_n atomic nuclei with electric charges $Z_\alpha e$ and masses M_α ($\alpha = 1, 2, \ldots, N_n$) and N_e electrons with charges $-e$ and masses m. Neglecting the finite structure of atomic nuclei, the Hamiltonian of the molecule can be written as

$$\hat{H} = \hat{H}_n + \hat{H}_e + V_{ne}, \tag{9.28}$$

where

$$\hat{H}_n = \sum_{\alpha=1}^{N_n} -\frac{\hbar^2}{2M_\alpha} \nabla_\alpha^2 + \frac{1}{2} \sum_{\substack{\alpha,\beta=1 \\ \alpha \neq \beta}}^{N_n} \frac{Z_\alpha Z_\beta e^2}{4\pi\varepsilon_0} \frac{1}{|\mathbf{R}_\alpha - \mathbf{R}_\beta|} \tag{9.29}$$

is the Hamiltonian of the atomic nuclei, with $\mathbf{R}_\alpha$ the position of the α-th nucleus,

$$\hat{H}_e = \sum_{i=1}^{N_e} -\frac{\hbar^2}{2m} \nabla_i^2 + \frac{1}{2} \sum_{\substack{i,j=1 \\ i \neq j}}^{N_e} \frac{e^2}{4\pi\varepsilon_0} \frac{1}{|\mathbf{r}_i - \mathbf{r}_j|} \tag{9.30}$$

is the Hamiltonian of the electrons, with $\mathbf{r}_i$ the position of the i-th electron, and

$$V_{ne} = -\sum_{\alpha,i=1}^{N_n, N_e} \frac{Z_\alpha e^2}{4\pi\varepsilon_0} \frac{1}{|\mathbf{R}_\alpha - \mathbf{r}_i|} \tag{9.31}$$

is the potential energy of the Coulomb interaction between atomic nuclei and electrons.

It is clear that the computation of the ground-state energy and the many-body wavefunction of an average-size molecule is a formidable task. For instance, the benzene molecule (C_6H_6) consists of 12 atomic nuclei and 42 electrons, and this means that its many-body wavefunction has $(12 + 42) \times 3 = 162$ variables: the spatial coordinates of the electrons and the nuclei. The exact many-body Schrodinger equation for the ground-state is given by

$$\hat{H}\Psi(R, r) = E \Psi(R, r), \tag{9.32}$$

where $\Psi(R, r) = \Psi(\mathbf{R}_1, \ldots, \mathbf{R}_{N_n}, \mathbf{r}_1, \ldots, \mathbf{r}_{N_e})$ is the ground-state wavefunction, with $R = (\mathbf{R}_1, \ldots, \mathbf{R}_{N_n})$ and $r = (\mathbf{r}_1, \ldots, \mathbf{r}_{N_e})$ multi-vectors for nuclear and electronic coordinates respectively.

In 1927 Max Born and Julius Robert Oppenheimer suggested a reliable approximation to treat this problem. Their approach is based on the separation of the fast electron dynamics from the slow motion of the nuclei. In the so-called Born-Oppenheimer approximation the many-body wave function of the molecule is factorized as

$$\Psi(R, r) = \Psi_e(r; R)\, \Psi_n(R)\,, \tag{9.33}$$

where $\Psi_n(R)$ is the nuclear wavefunction and $\Psi_e(r; R)$ is the electronic wavefunction, which depends also on nuclear coordinates. We do not analyze here the consequences of this factorization. However, we stress that Born-Oppenheimer approximation is crucial in any context where there is more than a single atom around, which includes atomic gases, clusters, crystals, and many other physical systems.

9.4   The Hartree-Fock Method

In 1927 by Douglas Hartree and Vladimir Fock used the variational principle to develop a powerful method for the study of interacting identical particles. We will analyze this variational method in the following subsections.

9.4.1   Hartree for Bosons

In the case of N identical interacting bosons the Hartree approximation is simply given by

$$\Psi(x_1, x_2, \ldots, x_N) = \phi(x_1)\, \phi(x_2)\, \ldots\, \phi(x_N)\,, \tag{9.34}$$

where the single-particle wavefunction $\phi(x)$ is unknown and it must be determined in a self-consistent way. Notice that, as previously stressed, this factorization implies that all particles belong to the same single-particle state, i.e. we are supposing that the interacting system is a pure Bose-Einstein condensate. This is a quite strong assumption, that is however reliable in the description of ultracold and dilute gases made of bosonic alkali-metal atoms (in 2001 Eric Cornell, Carl Weiman, and Wolfang Ketterle got the Nobel Prize in Physics for their experiments with these quantum gases), and which must be relaxed in the case of strongly-interacting bosonic systems (like superfluid ^{4}He). In the variational spirit of the Hartree approach the unknown wavefunction $\phi(x)$ is determined by minimizing the expectation value of the total Hamiltonian, given by

$$\langle \Psi | \hat{H} | \Psi \rangle = \int dx_1 dx_2 \ldots dx_N \Psi^*(x_1, x_2, \ldots, x_N) \hat{H} \Psi(x_1, x_2, \ldots, x_N)\,, \tag{9.35}$$

with respect to $\phi(x)$. In fact, by using Eq. (9.27) one finds immediately

$$\langle \Psi | \hat{H} | \Psi \rangle = N \int dx\, \phi^*(x)\hat{h}(x)\phi(x) + \frac{1}{2}N(N-1) \int dx\, dx'\, |\phi(x)|^2\, V(x, x')|\phi(x')|^2\,, \tag{9.36}$$

which is a nonlinear energy functional of the single-particle wavefunction $\phi(x)$. It is called single-orbital Hartree functional for bosons. In this functional the first term is related to the single-particle hamiltonian $\hat{h}(x)$ while the second term is related to the inter-particle interaction potential $V(x, x')$. We minimize this functional with the following constraint due to the normalization

$$\int dx \, |\phi(x)|^2 = 1 . \tag{9.37}$$

We get immediately the so-called Hartree equation for bosons

$$\left[\hat{h}(x) + U_{mf}(x)\right] \phi(x) = \epsilon \, \phi(x) , \tag{9.38}$$

where the mean-field potential $U_{mf}(x)$ reads

$$U_{mf}(x) = (N - 1) \int dx' \, V(x, x') \, |\phi(x')|^2 \tag{9.39}$$

and ϵ is the Lagrange multiplier fixed by the normalization. It is important to observe that the mean-field potential $U_{mf}(x)$ depends on $\phi(x)$ and it must be obtained self-consistently. In other words, the Hartree equation of bosons is a integro-differential nonlinear Schrödinger equation whose nonlinear term gives the mean-field potential of the system.

In the case of spinless bosons, where $|x\rangle = |\mathbf{r}\rangle$, given the local bosonic density

$$\rho(\mathbf{r}) = N |\phi(\mathbf{r})|^2 , \tag{9.40}$$

under the assumption of a large number N of particles the Hartree variational energy reads

$$\langle \Psi | \hat{H} | \Psi \rangle = N \int d^3 \mathbf{r} \, \phi^*(\mathbf{r}) \left[-\frac{\hbar^2}{2m} \nabla^2 + U(\mathbf{r}) \right] \phi(\mathbf{r}) + \frac{1}{2} \int d^3 \mathbf{r} \, d^3 \mathbf{r}' \, \rho(\mathbf{r}) V(\mathbf{r} - \mathbf{r}') \rho(\mathbf{r}') \tag{9.41}$$

while the Hartree equation becomes

$$\left[-\frac{\hbar^2}{2m} \nabla^2 + U(\mathbf{r}) + \int d^3 \mathbf{r}' \, V(\mathbf{r} - \mathbf{r}') \, \rho(\mathbf{r}') \right] \phi(\mathbf{r}) = \epsilon \, \phi(\mathbf{r}) . \tag{9.42}$$

To conclude this subsection, we observe that in the case of a contact inter-particle potential, i.e.

$$V(\mathbf{r} - \mathbf{r}') = g \, \delta(\mathbf{r} - \mathbf{r}') , \tag{9.43}$$

with the strength g given by

$$g = \int V(\mathbf{r}) \, d^3 \mathbf{r} , \tag{9.44}$$

the previous Hartree variational energy becomes

$$\langle \Psi | \hat{H} | \Psi \rangle = N \int d^3 \mathbf{r} \, \phi^*(\mathbf{r}) \left[-\frac{\hbar^2}{2m} \nabla^2 + U(\mathbf{r}) \right] \phi(\mathbf{r}) + \frac{g}{2} \int d^3 \mathbf{r} \, \rho(\mathbf{r})^2 \quad (9.45)$$

and the corresponding Hartee equation reads

$$\left[-\frac{\hbar^2}{2m} \nabla^2 + U(\mathbf{r}) + g \, \rho(\mathbf{r}) \right] \phi(\mathbf{r}) = \epsilon \, \phi(\mathbf{r}) \, , \quad (9.46)$$

which is the so-called Gross-Pitaevskii equation, deduced in 1961 by Eugene Gross and Lev Pitaevskii.

9.4.2   Hartree-Fock for Fermions

In the case of N identical interacting fermions, the approximation developed by Hartree and Fock is based   on the Slater determinant we have seen previously, namely

$$\Psi(x_1, x_2, \ldots, x_N) = \frac{1}{\sqrt{N!}} \begin{pmatrix} \phi_1(x_1) & \phi_1(x_2) & \ldots & \phi_1(x_N) \\ \phi_2(x_1) & \phi_2(x_2) & \ldots & \phi_2(x_N) \\ \ldots & \ldots & \ldots & \ldots \\ \phi_N(x_1) & \phi_N(x_2) & \ldots & \phi_N(x_N) \end{pmatrix} , \quad (9.47)$$

where now the single-particle wavefunctions $\phi_n(x)$ are unknown and they are determined with a variational procedure. In fact, in the Hartree-Fock approach the unknown wavefunctions $\phi_n(x)$ are obtained by minimizing the expectation value of the total Hamiltonian, given by

$$\langle \Psi | \hat{H} | \Psi \rangle = \int dx_1 \, dx_2 \ldots dx_N \, \Psi^*(x_1, x_2, \ldots, x_N) \hat{H} \Psi(x_1, x_2, \ldots, x_N) \, , \quad (9.48)$$

with respect to the N single-particle wavefunctions $\phi_n(x)$. By using Eq. (9.27) and after some tedious calculations one finds

$$\langle \Psi | \hat{H} | \Psi \rangle = \sum_{i=1}^{N} \int dx \, \phi_i^*(x) \hat{h}(x) \phi_i(x) + \frac{1}{2} \sum_{\substack{i,j=1 \\ i \neq j}}^{N} \left[\int dx \, dx' \, |\phi_i(x)|^2 \, V(x, x') |\phi_j(x')|^2 \right.$$
$$\left. - \int dx \, dx' \, \phi_i^*(x) \phi_j(x) V(x, x') \phi_j^*(x') \phi_i(x') \right] , \quad (9.49)$$

which is a nonlinear energy functional of the N single-particle wavefunctions $\phi_i(x)$. In this functional the first term is related to the single-particle Hamiltonian $\hat{h}(x)$ while the second and the third terms are related to the inter-particle interaction potential $V(x, x')$. The second term is called direct term of interaction and the third term is called exchange term of interaction. We minimize this functional with the following constraints due to the normalization

$$\int dx\, |\phi_i(x)|^2 = 1 , \qquad i = 1, 2, \ldots, N , \tag{9.50}$$

where, in the case of spin $1/2$ particles, one has

$$\phi_i(x) = \phi_i(\mathbf{r}, \sigma) = \tilde{\phi}_i(\mathbf{r}, \sigma)\, \chi_\sigma \tag{9.51}$$

with χ_σ the two-component spinor, and the integration over x means

$$\int dx = \int d^3\mathbf{r} \sum_{\sigma-\uparrow,\downarrow} , \tag{9.52}$$

such that

$$\int dx\, |\phi_i(x)|^2 = \int d^3\mathbf{r} \sum_{\sigma=\uparrow,\downarrow} |\phi_i(\mathbf{r}, \sigma)|^2 = \int d^3\mathbf{r} \sum_{\sigma-\uparrow,\downarrow} |\tilde{\phi}_i(\mathbf{r}, \sigma)|^2 , \tag{9.53}$$

because $\chi_\sigma^* \chi_\sigma = 1$ and more generally $\chi_\sigma^* \chi_{\sigma'} = \delta_{\sigma,\sigma'}$.

After minimization of the energy functional we get the so-called Hartree-Fock equations

$$\left[\hat{h}(x) + \hat{U}_{mf}(x)\right] \phi_i(x) = \epsilon_i\, \phi_i(x) \tag{9.54}$$

where ϵ_i are the Lagrange multipliers fixed by the normalization and $\hat{U}_{mf}$ is a nonlocal mean-field operator. This nonlocal operator is given by

$$\hat{U}_{mf}(x)\, \phi_i(x) = U_d(x)\, \phi_i(x) - \sum_{j=1}^{N} U_x^{ji}(x)\phi_j(x) , \tag{9.55}$$

where the direct mean-field potential $U_d(x)$ reads

$$U_d(x) = \sum_{\substack{j=1 \\ j\neq i}}^{N} \int dx'\, V(x, x')\, |\phi_j(x')|^2 , \tag{9.56}$$

while the exchange mean-field potential $U_x^{ji}(x)$ is instead

$$U_x^{ji}(x) = \int dx'\, \phi_j^*(x')\, V(x, x')\, \phi_i(x') . \tag{9.57}$$

If one neglects the exchange term, as done by Hartree in his original derivation, the so-called Hartree equations

$$\left[\hat{h}(x) + U_d(x)\right] \phi_i(x) = \epsilon_i\, \phi_i(x)\,, \tag{9.58}$$

are immediately derived. It is clearly much simpler to solve the Hartree equations instead of the Hartree-Fock ones. For this reason, in many applications the latter are often used. In the case of spin $1/2$ fermions, given the local fermionic density

$$\rho(\mathbf{r}) = \sum_{\sigma=\uparrow,\downarrow} \sum_{i=1}^{N} |\phi_i(\mathbf{r}, \sigma)|^2 = \sum_{\sigma=\uparrow,\downarrow} \rho(\mathbf{r}, \sigma)\,, \tag{9.59}$$

under the assumption of a large number N of particles the Hartree (direct) variational energy reads

$$E_D = \sum_{i=1}^{N} \sum_{\sigma=\uparrow,\downarrow} \int d^3\mathbf{r}\, \phi_i^*(\mathbf{r}, \sigma)\left[-\frac{\hbar^2}{2m}\nabla^2 + U(\mathbf{r})\right]\phi_i(\mathbf{r}, \sigma)$$
$$+ \frac{1}{2}\sum_{\sigma,\sigma'=\uparrow,\downarrow} \int d^3\mathbf{r}\, d^3\mathbf{r}'\, \rho(\mathbf{r}, \sigma)V(\mathbf{r} - \mathbf{r}')\rho(\mathbf{r}', \sigma') \tag{9.60}$$

and the corresponding Hartree equation becomes

$$\left[-\frac{\hbar^2}{2m}\nabla^2 + U(\mathbf{r}) + \sum_{\sigma'=\uparrow,\downarrow} \int d^3\mathbf{r}'\, V(\mathbf{r} - \mathbf{r}')\, \rho(\mathbf{r}', \sigma')\right] \phi_i(\mathbf{r}, \sigma) = \epsilon_i\, \phi_i(\mathbf{r}, \sigma)\,. \tag{9.61}$$

The Hartree-Fock variational energy is slightly more complex because it includes also the exchange energy, given by

$$E_X = -\frac{1}{2}\sum_{\substack{i,j=1 \\ i\neq j}}^{N} \sum_{\sigma=\uparrow,\downarrow} \int d^3\mathbf{r}\, d^3\mathbf{r}'\, \phi_i^*(\mathbf{r}, \sigma)\phi_i(\mathbf{r}', \sigma)V(\mathbf{r} - \mathbf{r}')\phi_j^*(\mathbf{r}', \sigma)\phi_j(\mathbf{r}, \sigma)\,. \tag{9.62}$$

Notice that in the exchange energy all the terms with opposite spins are zero due to the scalar product of spinors: $\chi_\sigma^* \chi_{\sigma'} = \delta_{\sigma,\sigma'}$. The existence of this exchange energy E_X is a direct consequence of the anti-symmetry of the many-body wave function, namely a consequence of the fermionic nature of the particles we are considering. Historically, this term E_X was obtained by Vladimir Fock to correct the first derivation of Douglas Hartree who used a not anti-symmetrized many-body wavefunction.

To conclude this subsection, we observe that in the case of a contact inter-particle potential, i.e.

$$V(\mathbf{r} - \mathbf{r}') = g\, \delta(\mathbf{r} - \mathbf{r}')\,, \tag{9.63}$$

the Hartee-Fock (direct plus exchange) variational energy reads

$$E = \sum_{i=1}^{N} \sum_{\sigma=\uparrow,\downarrow} \int d^3\mathbf{r}\, \phi_i^*(\mathbf{r},\sigma) \left[-\frac{\hbar^2}{2m}\nabla^2 + U(\mathbf{r}) \right] \phi_i(\mathbf{r},\sigma) \tag{9.64}$$

$$+ \frac{g}{2} \sum_{\substack{i,j=1 \\ i\neq j}}^{N} \sum_{\sigma,\sigma'=\uparrow,\downarrow} \int d^3\mathbf{r}\, \left[\, |\phi_i(\mathbf{r},\sigma)|^2 |\phi_j(\mathbf{r},\sigma')|^2 - |\phi_i(\mathbf{r},\sigma)|^2 |\phi_j(\mathbf{r},\sigma)|^2 \delta_{\sigma,\sigma'} \,\right].$$

It follows immediately that identical spin-polarized fermions with contact interaction are effectively non-interacting because in this case the interaction terms of direct and exchange energy exactly compensate to zero.

Further Reading

A classic book on the many-body quantum problem is:
Fetter, A.L., Walecka, J.D.: Quantum Theory of Many-Particle Systems. Dover Publications (2003)
For the many-electron atom and the Hartree-Fock method:
Bransden, B.H., Joachain, C.J.: Physics of Atoms and Molecules. Prentice Hall (2003)
Lipparini, E.: Modern Many-Particle Physics: Atomic Gases, Quantum Dots and Quantum Fluids. World Scientific (2003)

Chapter 10
Quantum Statistical Mechanics

In this chapter we discuss elements of quantum statistical mechanics. In the first chapter we have seen that statistical mechanics aims to describe macroscopic properties of complex systems starting from their microscopic components by using statistical averages. Quantum statistical mechanics is more general than classical statistical mechanics and it reproduces all the results of thermodynamics. In general, quantum statistical machanics reduces to classical statistical mechanics in the high-temperature regime.

10.1 Quantum Statistical Ensembles

Here we discuss only quantum systems at thermal equilibrium and consider a many-body quantum system of identical particles characterized by the Hamiltonian $\hat{H}$ such that

$$\hat{H}|E_i^{(N)}\rangle = E_i^{(N)}|E_i^{(N)}\rangle, \tag{10.1}$$

where $|E_i^{(N)}\rangle$ are the eigenstates of $\hat{H}$ for a fixed number N of identical particles and E_i^N are the corresponding eigenenergies.

10.1.1 Quantum Microcanonical Ensemble

In the microcanonical ensemble the quantum many-body system in a volume V has a fixed number N of particles and also a fixed energy E. In this case the Hamiltonian $\hat{H}$ admits the spectral decomposition

© The Author(s), under exclusive license to Springer Nature Switzerland AG 2022
L. Salasnich, *Modern Physics*, UNITEXT for Physics,
https://doi.org/10.1007/978-3-030-93743-0_10

$$\hat{H} = \sum_i E_i^{(N)} |E_i^{(N)}\rangle \langle E_i^{(N)}|, \tag{10.2}$$

and one defines the microcanonical density operator as

$$\hat{\rho} = W_0 \, \delta\left(E - \hat{H}\right), \tag{10.3}$$

where $\delta(x)$ is the Dirac delta function and W_0 is an arbitrary constant with the units of an energy such that the density operator is adimensional. This microcanonical density operator $\hat{\rho}$ has the spectral decomposition

$$\hat{\rho} = \sum_i W_0 \, \delta\left(E - E_i^{(N)}\right) |E_i^{(N)}\rangle \langle E_i^{(N)}|. \tag{10.4}$$

The key quantity in the microcanonical ensemble is the density of states (or microcanonical volume) W given by

$$W = Tr[\hat{\rho}] = Tr[W_0 \, \delta\left(E - \hat{H}\right)], \tag{10.5}$$

namely

$$W = \sum_i W_0 \, \delta(E - E_i^{(N)}). \tag{10.6}$$

The ensemble average of an observable described by the self-adjunct operator $\hat{A}$ is defined as

$$\langle \hat{A} \rangle = \frac{Tr[\hat{A} \, \hat{\rho}]}{Tr[\hat{\rho}]} = \frac{1}{W} \sum_i A_{ii}^{(N)} \, W_0 \, \delta\left(E, E_i^{(N)}\right), \tag{10.7}$$

where $A_{ii}^{(N)} = \langle E_i^{(N)} | \hat{A} | E_i^{(N)} \rangle$. As in the case of classical statistical mechanics, the connection with equilibrium thermodynamics is given by the formula

$$S = k_B \, \ln(W), \tag{10.8}$$

which introduces the entropy S as a function of energy E, volume V and number N of particles.

10.1.2 Quantum Canonical Ensemble

In the canonical ensemble the quantum system in a volume V has a fixed number N of particles and a fixed temperature T. In this case one defines the canonical density operator as

$$\hat{\rho} = e^{-\beta\hat{H}},$$ (10.9)

with $\beta = 1/(k_B T)$. Here $\hat{\rho}$ has the spectral decomposition

$$\hat{\rho} = \sum_i e^{-\beta E_i^{(N)}} |E_i^{(N)}\rangle\langle E_i^{(N)}|.$$ (10.10)

The key quantity in the canonical ensemble is the canonical partition function (or canonical volume) $\mathcal{Z}_N$ given by

$$\mathcal{Z}_N = Tr[\hat{\rho}] = Tr[e^{-\beta\hat{H}}],$$ (10.11)

namely

$$\mathcal{Z}_N = \sum_i e^{-\beta E_i^{(N)}}.$$ (10.12)

The ensemble average of an observable $\hat{A}$ is defined as

$$\langle\hat{A}\rangle = \frac{Tr[\hat{A}\,\hat{\rho}]}{Tr[\hat{\rho}]} = \frac{1}{\mathcal{Z}_N}\sum_i A_{ii}^{(N)}\, e^{-\beta E_i^{(N)}},$$ (10.13)

where $A_{ii}^{(N)} = \langle E_i^{(N)}|\hat{A}|E_i^{(N)}\rangle$. Notice that the definition of canonical-ensemble average is the same of the microcanonical-ensemble average but the density of state $\hat{\rho}$ is different in the two ensembles. The connection with equilibrium thermodynamics is given by the formula

$$\mathcal{Z}_N = e^{-\beta F},$$ (10.14)

which introduces the Helmholtz free energy F as a function of temperature T, volume V and number N of particles.

10.1.3 Quantum Grand Canonical Ensemble

In the grand canonical ensemble the quantum system in a volume V has a fixed temperature T and a fixed chemical potential μ. In this case the Hamiltonian $\hat{H}$ has the spectral decomposition

$$\hat{H} = \sum_{N=0}^{\infty}\sum_i E_i^{(N)} |E_i^{(N)}\rangle\langle E_i^{(N)}|,$$ (10.15)

which is a generalization of Eq. (10.2), and one introduces the total number operator $\hat{N}$ such that

$$\hat{N}|E_i^{(N)}\rangle = N|E_i^{(N)}\rangle, \tag{10.16}$$

and consequently $\hat{N}$ has the spectral decomposition

$$\hat{N} = \sum_{N=0}^{\infty} \sum_i N|E_i^{(N)}\rangle\langle E_i^{(N)}|. \tag{10.17}$$

For the grand canonical ensemble one defines the grand canonical density operator as

$$\hat{\rho} = e^{-\beta(\hat{H}-\mu\hat{N})}, \tag{10.18}$$

with $\beta = 1/(k_B T)$ and μ the chemical potential. Here $\hat{\rho}$ has the spectral decomposition

$$\hat{\rho} = \sum_{N=0}^{\infty} \sum_i e^{-\beta(E_i^{(N)}-\mu N)}|E_i^{(N)}\rangle\langle E_i^{(N)}| = \sum_{N=0}^{\infty} z^N \sum_i e^{-\beta E_i^{(N)}}|E_i^{(N)}\rangle\langle E_i^{(N)}|, \tag{10.19}$$

where $z = e^{\beta\mu}$ is the fugacity. The key quantity in the grand canonical ensemble is the grand canonical partition function (or grand canonical volume) $\mathcal{Z}$ given by

$$\mathcal{Z} = Tr[\hat{\rho}] = Tr[e^{-\beta(\hat{H}-\mu\hat{N})}], \tag{10.20}$$

namely

$$\mathcal{Z} = \sum_{N=0}^{\infty} \sum_i e^{-\beta(E_i^{(N)}-\mu N)} = \sum_{N=0}^{\infty} z^N \, \mathcal{Z}_N. \tag{10.21}$$

The ensemble average of an observable $\hat{A}$ is defined as

$$\langle\hat{A}\rangle = \frac{Tr[\hat{A}\,\hat{\rho}]}{Tr[\hat{\rho}]} = \frac{1}{\mathcal{Z}} \sum_{N=0}^{\infty} \sum_i A_{ii}^{(N)} \, e^{-(\beta E_i^{(N)}-\mu N)}, \tag{10.22}$$

where $A_{ii}^{(N)} = \langle E_i^{(N)}|\hat{A}|E_i^{(N)}\rangle$. Notice that the definition of grand canonical-ensemble average is the same of both microcanonical-ensemble average and canonical-enseble average but the density of state $\hat{\rho}$ is different in the three ensembles. Also here, the connection with equilibrium thermodynamics is given by the same formula of the classical case, i.e.

$$\mathcal{Z} = e^{-\beta\Omega}, \tag{10.23}$$

which introduces the grand potential Ω as a function of temperature T, volume V and chemical potential μ.

To conclude this section we observe that in the grand canonical ensemble, instead of working with eigenstates $|E_i^{(N)}\rangle$ of $\hat{H}$ at fixed number N of particles, one can work with multi-mode Fock states

$$|n_0\, n_1\, n_2 \ldots n_\infty\rangle = |n_0\rangle \otimes |n_1\rangle \otimes |n_2\rangle \otimes \ldots \otimes |n_\infty\rangle, \qquad (10.24)$$

where $|n_\alpha\rangle$ is the single-mode Fock state which describes n_α particles in the single-mode state $|\alpha\rangle$ with $\alpha = 0, 1, 2, \ldots$. The trace Tr which appears in Eq. (10.20) is indeed independent on the basis representation.

10.2   Bosons and Fermions at Finite Temperature

Let us consider the non-interacting matter field in thermal equilibrium with a bath at the temperature T. The relevant quantity to calculate all thermodynamical properties of the system is the grand-canonical partition function $\mathcal{Z}$, given by

$$\mathcal{Z} = Tr[e^{-\beta(\hat{H}-\mu\hat{N})}] \qquad (10.25)$$

where $\beta = 1/(k_B T)$ with k_B the Boltzmann constant,

$$\hat{H} = \sum_\alpha \epsilon_\alpha \hat{N}_\alpha, \qquad (10.26)$$

is the quantum Hamiltonian,

$$\hat{N} = \sum_\alpha \hat{N}_\alpha \qquad (10.27)$$

is total number operator, and μ is the chemical potential, fixed by the conservation of the average particle number. This implies that

$$\mathcal{Z} = \sum_{\{n_\alpha\}} \langle \ldots n_\alpha \ldots | e^{-\beta(\hat{H}-\mu\hat{N})} | \ldots n_\alpha \ldots \rangle = \sum_{\{n_\alpha\}} \langle \ldots n_\alpha \ldots | e^{-\beta\sum_\alpha(\epsilon_\alpha-\mu)\hat{N}_\alpha} | \ldots n_\alpha \ldots \rangle$$

$$= \sum_{\{n_\alpha\}} e^{-\beta\sum_\alpha(\epsilon_\alpha-\mu)n_\alpha} = \sum_{\{n_\alpha\}} \prod_\alpha e^{-\beta(\epsilon_\alpha-\mu)n_\alpha} = \prod_\alpha \sum_{n_\alpha} e^{-\beta(\epsilon_\alpha-\mu)n_\alpha}$$

$$= \prod_\alpha \sum_{n=0}^{\infty} e^{-\beta(\epsilon_\alpha-\mu)n} = \prod_\alpha \frac{1}{1-e^{-\beta(\epsilon_\alpha-\mu)}} \qquad \text{for bosons} \qquad (10.28)$$

$$= \prod_\alpha \sum_{n=0}^{1} e^{-\beta(\epsilon_\alpha-\mu)n} = \prod_\alpha \left(1+e^{-\beta(\epsilon_\alpha-\mu)}\right) \qquad \text{for fermions} \qquad (10.29)$$

Quantum statistical mechanics dictates that the thermal average of any operator $\hat{A}$ is obtained as

$$\langle \hat{A} \rangle = \frac{1}{\mathcal{Z}} Tr[\hat{A}\, e^{-\beta(\hat{H}-\mu\hat{N})}]. \qquad (10.30)$$

Let us suppose that $\hat{A} = \hat{H}' = \hat{H} - \mu\hat{N}$, it is then quite easy to show that

$$\langle \hat{H}' \rangle = \frac{1}{\mathcal{Z}} Tr[(\hat{H} - \mu\hat{N}) e^{-\beta(\hat{H}-\mu\hat{N})}] = -\frac{\partial}{\partial \beta} \ln \left(Tr[e^{-\beta(\hat{H}-\mu\hat{N})}] \right) = -\frac{\partial}{\partial \beta} \ln(\mathcal{Z}).$$
(10.31)

By using Eq. (10.28) or Eq. (10.29) we immediately obtain

$$\ln(\mathcal{Z}) = \mp \sum_{\alpha} \ln \left(1 \mp e^{-\beta(\epsilon_\alpha - \mu)} \right),$$
(10.32)

where $-$ is for bosons and $+$ for fermions, and finally from Eq. (10.31) we get

$$\langle \hat{H} \rangle = \sum_{\alpha} \epsilon_\alpha \langle \hat{N}_\alpha \rangle_T,$$
(10.33)

with

$$\langle \hat{N} \rangle = \sum_{\alpha} \frac{1}{e^{\beta(\epsilon_\alpha - \mu)} \mp 1}.$$
(10.34)

Notice that the zero-temperature limit, i.e. $\beta \to \infty$, for fermions we have

$$\langle \hat{N} \rangle = \sum_{\alpha} \Theta(\mu - \epsilon_\alpha),$$
(10.35)

where $\Theta(x)$ is the Heaviside step function and the chemical potential μ at zero temperature is nothing but the Fermi energy ϵ_F, i.e. $\epsilon_F = \mu(T = 0)$. Instead, in the high-temperature regime, where $\beta(\epsilon_\alpha - \mu) \gg 1$, Eq. (10.34) becomes

$$\langle \hat{N} \rangle = \sum_{\alpha} e^{-\beta(\epsilon_\alpha - \mu)},$$
(10.36)

that is exactly the Maxwell-Boltzmann distribution. Thus, when the temperature T is such that also the highly excited energy levels ϵ_α are occupied, then both Bose-Einstein and Fermi-Dirac distributions reduce to the Maxwell-Boltzmann one.

10.2.1   Gas of Photons at Thermal Equilibrium

The chemical potential μ of a many-body system is the energy required to add a particle to the system. The minimal energy to create a particle of mass m from the vacuum is nothing else than its rest energy mc^2. Thus, for a gas of non-interacting photons we have $\mu = 0$ and consequently the number of photons is not fixed. This implies that

$$\langle \hat{H} \rangle = \sum_{k} \sum_{s} \frac{\hbar \omega_k}{e^{\beta \hbar \omega_k} - 1} = \sum_{k} \sum_{s} \hbar \omega_k \langle \hat{N}_{ks} \rangle. \tag{10.37}$$

In the continuum limit, where

$$\sum_{k} \rightarrow V \int \frac{d^3 k}{(2\pi)^3}, \tag{10.38}$$

with V the volume, and taking into account that $\omega_k = ck$, one can write the energy density $\mathcal{E} = \langle \hat{H} \rangle_T / V$ as

$$\mathcal{E} = 2 \int \frac{d^3 \mathbf{k}}{(2\pi)^3} \frac{c\hbar k}{e^{\beta c \hbar k} - 1} = \frac{c\hbar}{\pi^2} \int_0^\infty dk \frac{k^3}{e^{\beta c \hbar k} - 1}, \tag{10.39}$$

where the factor 2 is due to the two possible polarizations ($s = 1, 2$). By using $\omega = ck$ instead of k as integration variable one gets

$$\mathcal{E} = \frac{\hbar}{\pi^2 c^3} \int_0^\infty d\omega \frac{\omega^3}{e^{\beta \hbar \omega} - 1} = \int_0^\infty d\omega \, \rho(\omega), \tag{10.40}$$

where

$$\rho(\omega) = \frac{\hbar}{\pi^2 c^3} \frac{\omega^3}{e^{\beta \hbar \omega} - 1} \tag{10.41}$$

is the energy density per frequency, i.e. the familiar formula of the black-body radiation, obtained for the first time in 1900 by Max Planck. The previous integral can be explicitly calculated and it gives

$$\mathcal{E} = \frac{\pi^2 k_B^4}{15 c^3 \hbar^3} T^4, \tag{10.42}$$

which is nothing but the Stefan-Boltzmann law. In an similar way one determines the average number density of photons:

$$n = \frac{\langle \hat{N} \rangle_T}{V} = \frac{1}{\pi^2 c^3} \int_0^\infty d\omega \frac{\omega^2}{e^{\beta \hbar \omega} - 1} = \frac{2\zeta(3) k_B^3}{\pi^2 c^3 \hbar^3} T^3. \tag{10.43}$$

where $\zeta(3) \simeq 1.202$. Notice that both energy density $\mathcal{E}$ and number density n of photons go to zero as the temperature T goes to zero. To conclude this section, we stress that these results are obtained at thermal equilibrium and under the condition of a vanishing chemical potential, meaning that the number of photons is not conserved when the temperature is varied.

10.2.2 Gas of Massive Bosons at Thermal Equlibrium

For a gas of non-interacting bosons, which can occupy the single-particle energy
levels ϵ_α of the single-particle quantum states $|\alpha\rangle$, we have found that the thermal
average of the internal energy is given by

$$\bar{E} = \sum_\alpha \epsilon_\alpha \, \bar{N}_\alpha, \tag{10.44}$$

where

$$\bar{N}_\alpha = \frac{1}{e^{\beta(\epsilon_\alpha - \mu)} - 1} \tag{10.45}$$

is the termal average of the number of bosons in the single-particle quantum state
$|\alpha\rangle$. That is the Bose-Einstein distribution. Here, to simplify a bit the notation, we
have used the bar over the symbol to represent the thermal average, i.e. $\bar{E} = \langle \hat{H} \rangle$
and $\bar{N}_\alpha = \langle \hat{N}_\alpha \rangle$.

The thermal average $\bar{N} = \langle \hat{N} \rangle$ of the total number of bosons then reads

$$\bar{N} = \sum_\alpha \bar{N}_\alpha = \sum_\alpha \frac{1}{e^{\beta(\epsilon_\alpha - \mu)} - 1}. \tag{10.46}$$

At fixed temperature $T = \frac{1}{k_B \beta}$, $\bar{N}$ is fully determined by the chemical potential μ.

Let us suppose for simplicity that the set of single-particle energy levels ϵ_α is
given by

$$\epsilon_0, \, \epsilon_1, \, \epsilon_2, \, \epsilon_3, \, ... \tag{10.47}$$

where

$$\epsilon_0 < \epsilon_1 \leq \epsilon_2 \leq \epsilon_3 \leq \, ... \tag{10.48}$$

Clearly it must be

$$\mu < \epsilon_0 \tag{10.49}$$

to avoid divergences in the Bose-Einstein distributions of each

$$\bar{N}_\alpha = \frac{1}{e^{\beta(\epsilon_\alpha - \mu)} - 1}. \tag{10.50}$$

Moreover, as $\mu \to \epsilon_0$ the distribution

$$\bar{N}_0 = \frac{1}{e^{\beta(\epsilon_0 - \mu)} - 1} \tag{10.51}$$

becomes very large: under this condition we have Bose-Einstein condensation, i.e.
a macrosopic number of bosons in the lowest single-particle energy level ϵ_0.

In the presence of Bose-Einstein condensation (BEC) it is useful to write the total number of bosons as follows

$$\bar{N} = \bar{N}_0 + \sum_{\alpha \neq 0} \bar{N}_\alpha = \frac{1}{e^{\beta(\epsilon_0 - \mu)} - 1} + \sum_{\alpha \neq 0} \frac{1}{e^{\beta(\epsilon_\alpha - \mu)} - 1}. \tag{10.52}$$

The exact condensate fraction is defined as

$$\frac{\bar{N}_0}{\bar{N}} = 1 - \frac{\sum_{\alpha \neq 0} \frac{1}{e^{\beta(\epsilon_\alpha - \mu)} - 1}}{\sum_{\alpha} \frac{1}{e^{\beta(\epsilon_\alpha - \mu)} - 1}}. \tag{10.53}$$

Following Einstein (1924), instead of using the exact formula of $\bar{N}_0$ we assume that N_0 is unknown but we also set

$$\mu = \epsilon_0 \tag{10.54}$$

in the BEC phase (i.e. when $N_0 > 0$). In this way, in the BEC phase we find

$$\bar{N} = N_0 + \sum_{\alpha \neq 0} \frac{1}{e^{\beta(\epsilon_\alpha - \epsilon_0)} - 1}. \tag{10.55}$$

At the critical temperature T_{BEC} we have $\bar{N}_0 = 0$ and consequently

$$\bar{N} = \sum_{\alpha \neq 0} \frac{1}{e^{(\epsilon_\alpha - \epsilon_0)/(k_B T_{BEC})} - 1}, \tag{10.56}$$

while for $T < T_{BEC}$ we have

$$\bar{N} = \bar{N}_0 + \sum_{\alpha \neq 0} \frac{1}{e^{(\epsilon_\alpha - \epsilon_0)/(k_B T)} - 1}. \tag{10.57}$$

Then the condensate fraction reads

$$\frac{\bar{N}_0}{\bar{N}} = 1 - \frac{\sum_{\alpha \neq 0} \frac{1}{e^{(\epsilon_\alpha - \mu)/(k_B T)} - 1}}{\sum_{\alpha \neq 0} \frac{1}{e^{(\epsilon_\alpha - \mu)/(k_B T_{BEC})} - 1}}. \tag{10.58}$$

Let us now consider a Bose gas of atoms with mass m in a cubic box of volume L^D. The single-particle energy spectrum is

$$\epsilon_{\mathbf{q}} = \frac{\hbar^2 q^2}{2m}. \tag{10.59}$$

In the continuum limit

$$\sum_{\mathbf{q}} \rightarrow L^D \int \frac{d^D\mathbf{q}}{(2\pi)^D} \tag{10.60}$$

and the total number density $\bar{n} = \bar{N}/L^D$ in the BEC phase is given by

$$\bar{n} = \bar{n}_0 + \int \frac{d^D\mathbf{q}}{(2\pi)^D} \frac{1}{e^{\frac{\hbar^2 q^2}{2mk_BT}} - 1}, \tag{10.61}$$

where $\bar{n}_0 = \bar{N}_0/L^D$ is the condensate number density.

The critical temperature T_{BEC} of Bose-Einstein condensation is obtained setting $\bar{n}_0 = 0$ in the previous equation. In this way one finds

$$k_B T_{BEC} = \begin{cases} \frac{2\pi}{\zeta(3/2)^{2/3}} \frac{\hbar^2}{m} \bar{n}^{2/3} & \text{for } D = 3 \\ 0 & \text{for } D = 2 \\ \text{no solution} & \text{for } D = 1 \end{cases} \tag{10.62}$$

where $\zeta(x)$ is the Riemann zeta function. This result was extended to interacting systems by David Mermin and Herbert Wagner in 1966. The so-called Mermin-Wagner theorem states that there is no Bose-Einstein condensation at finite temperature in homogeneous systems with sufficiently short-range interactions in dimensions $D \leq 2$.

In the three-dimensional case ($D = 3$) from the equation

$$\bar{n} = \bar{n}_0 + \int \frac{d^3\mathbf{q}}{(2\pi)^3} \frac{1}{e^{\frac{\hbar^2 q^2}{2mk_BT}} - 1}, \tag{10.63}$$

we find

$$\bar{n} = \bar{n}_0 + \zeta(3/2) \left(\frac{mk_BT}{2\pi\hbar^2} \right)^{3/2}. \tag{10.64}$$

It follows that

$$\frac{\bar{n}_0}{\bar{n}} = 1 - \frac{\zeta(3/2) \left(\frac{mk_B}{2\pi\hbar^2} \right)^{3/2} T^{3/2}}{\bar{n}} = 1 - \frac{\zeta(3/2) \left(\frac{mk_B}{2\pi\hbar^2} \right)^{3/2} T^{3/2}}{\zeta(3/2) \left(\frac{mk_B}{2\pi\hbar^2} \right)^{3/2} T_{BEC}^{3/2}}. \tag{10.65}$$

Thus, the condensate fraction reads

$$\frac{\bar{n}_0}{\bar{n}} = 1 - \left(\frac{T}{T_{BEC}} \right)^{3/2}. \tag{10.66}$$

The critical temperature of Eq. (10.62) and this formula for the condensate fraction of non-interacting massive bosons were obtained in 1925 by Albert Einstein extending previous results derived by Satyendra Nath Bose for a gas a photons (massless bosons).

10.2.3 Gas of Non-interacting Fermions at Zero Temperature

A quite important physical system is the uniform gas of non-interacting fermions. It is indeed a good starting point for the description of all the real systems which have a finite interaction between fermions.

The non-interacting uniform Fermi gas is obtained setting to zero the confining potential, i.e.

$$U(\mathbf{r}) = 0, \tag{10.67}$$

and imposing periodicity conditions on the single-particle wavefunctions, which are plane waves with a spinor

$$\phi(x) = \frac{1}{\sqrt{V}} e^{i\mathbf{k}\cdot\mathbf{r}} \chi_\sigma, \tag{10.68}$$

where χ_σ is the spinor for spin-up and spin-down along a chosen z asis:

$$\chi_\uparrow = \begin{pmatrix} 1 \\ 0 \end{pmatrix}, \qquad \chi_\downarrow = \begin{pmatrix} 0 \\ 1 \end{pmatrix}. \tag{10.69}$$

At the boundaries of a cube having volume V and side L one has

$$e^{ik_x(x+L)} = e^{ik_x x}, \quad e^{ik_y(y+L)} = e^{ik_y y}, \quad e^{ik_z(z+L)} = e^{ik_z z}. \tag{10.70}$$

It follows that the linear momentum $\mathbf{k}$ can only take on the values

$$k_x = \frac{2\pi}{L} n_x, \quad k_y = \frac{2\pi}{L} n_y, \quad k_z = \frac{2\pi}{L} n_z, \tag{10.71}$$

where n_x, n_y, n_z are integer quantum numbers. The single-particle energies are given by

$$\epsilon_{\mathbf{k}} = \frac{\hbar^2 k^2}{2m} = \frac{\hbar^2}{2m} \frac{4\pi^2}{L^2} (n_x^2 + n_y^2 + n_z^2). \tag{10.72}$$

In the thermodynamic limit $L \to \infty$, the allowed values are closely spaced and one can use the continuum approximation

$$\sum_{n_x, n_y, n_z} \to \int dn_x \, dn_y \, dn_z, \tag{10.73}$$

which implies

$$\sum_{\mathbf{k}} \rightarrow \frac{L^3}{(2\pi)^3} \int d^3\mathbf{k} = V \int \frac{d^3\mathbf{k}}{(2\pi)^3}. \tag{10.74}$$

The total number N of fermionic particles is given by

$$N = \sum_{\sigma} \sum_{\mathbf{k}} \Theta\left(\epsilon_F - \epsilon_{\mathbf{k}}\right), \tag{10.75}$$

where the Heaviside step function $\Theta(x)$, such that $\Theta(x) = 0$ for $x < 0$ and $\Theta(x) = 1$ for $x > 0$, takes into account the fact that fermions are occupied only up to the Fermi energy ϵ_F, which is determined by fixing N. Notice that at finite temperature T the total number N of ideal fermions is instead obtained from the Fermi-Dirac distribution, namely

$$N = \sum_{\sigma} \sum_{\mathbf{k}} \frac{1}{e^{\beta(\epsilon_{\mathbf{k}} - \mu)} + 1}, \tag{10.76}$$

where $\beta = 1/(k_B T)$, with k_B the Boltzmann constant, and μ is the chemical potential of the system. In the limit $\beta \rightarrow +\infty$, i.e. for $T \rightarrow 0$, the Fermi-Dirac distribution becomes the Heaviside step function and μ is identified as the Fermi energy ϵ_F.

In the continuum limit and choosing spin $1/2$ fermions one finds

$$N = \sum_{\sigma=\uparrow,\downarrow} V \int \frac{d^3\mathbf{k}}{(2\pi)^3} \Theta\left(\epsilon_F - \frac{\hbar^2 k^2}{2m}\right), \tag{10.77}$$

from which one gets (the sum of spins gives simply a factor 2) the uniform density

$$\rho = \frac{N}{V} = \frac{1}{3\pi^2} \left(\frac{2m\epsilon_F}{\hbar^2}\right)^{3/2}. \tag{10.78}$$

The formula can be inverted giving the Fermi energy ϵ_F as a function of the density ρ, namely

$$\epsilon_F = \frac{\hbar^2}{2m} \left(3\pi^2 \rho\right)^{2/3}. \tag{10.79}$$

In many applications the Fermi energy ϵ_F is written as

$$\epsilon_F = \frac{\hbar^2 k_F^2}{2m}, \tag{10.80}$$

where k_F is the so-called Fermi wave-number, given by

$$k_F = \left(3\pi^2 \rho\right)^{1/3}. \tag{10.81}$$

The total energy E of the uniform and non-interacting Fermi system is given by

$$E = \sum_{\sigma} \sum_{\mathbf{k}} \epsilon_{\mathbf{k}} \, \Theta \left(\epsilon_F - \epsilon_{\mathbf{k}} \right), \tag{10.82}$$

and using again the continuum limit with spin $1/2$ fermions it becomes

$$E = \sum_{\sigma=\uparrow,\downarrow} V \int \frac{d^3 \mathbf{k}}{(2\pi)^3} \frac{\hbar^2 k^2}{2m} \, \Theta \left(\epsilon_F - \frac{\hbar^2 k^2}{2m} \right), \tag{10.83}$$

from which one gets the energy density

$$\mathcal{E} = \frac{E}{V} = \frac{3}{5} \rho \, \epsilon_F = \frac{3}{5} \frac{\hbar^2}{2m} \left(3\pi^2 \right)^{2/3} \rho^{5/3} \tag{10.84}$$

in terms of the Fermi energy ϵ_F and the uniform density ρ.

Further Reading

Two classic books where the formalism of quantum statistical mechanics is introduced are:
A.L. Fetter and J.D. Walecka, Quantum Theory of Many-Particle Systems (Dover Publications, 2003).
K. Huang, Statistical Mechanics (Wiley, 1987).
A mathematically oriented book on the same subject is:
W.C. Schieve and L.P. Horwitz, Quantum Statistical Mechanics (Cambridge Univ. Press, 2009).

Appendix A
Dirac Delta Function

The Heaviside Step Function

In 1880 the self-taught electrical scientist Oliver Heaviside introduced the following function

$$\Theta(x) = \begin{cases} 1 \text{ for } x > 0 \\ 0 \text{ for } x < 0 \end{cases} \tag{A.1}$$

which is now called Heaviside step function. This is a discontinous function, with a discontinuity of first kind (jump) at $x = 0$, which is often used in the context of the analysis of electric signals.

Moreover, it is important to stress that the Haviside step function appears also in the context of quantum statistical physics. In fact, the Fermi-Dirac function (or Fermi-Dirac distribution)

$$F_\beta(x) = \frac{1}{e^{\beta x} + 1}, \tag{A.2}$$

proposed in 1926 by Enrico Fermi and Paul Dirac to describe the quantum statistical distribution of electrons in metals, where $\beta = 1/(k_B T)$ is the inverse of the absolute temperature T (with k_B the Boltzmann constant) and $x = \epsilon - \mu$ is the energy ϵ of the electron with respect to the chemical potential μ, becomes the function $\Theta(-x)$ in the limit of very small temperature T, namely

$$\lim_{\beta \to +\infty} F_\beta(x) = \Theta(-x) = \begin{cases} 0 \text{ for } x > 0 \\ 1 \text{ for } x < 0 \end{cases}. \tag{A.3}$$

The Strange Function of Dirac

Inspired by the work of Heaviside, with the purpose of describing an extremely localized charge density, in 1930 Paul Dirac investigated the following "function"

$$\delta(x) = \begin{cases} +\infty & \text{for } x = 0 \\ 0 & \text{for } x \neq 0 \end{cases} \tag{A.4}$$

imposing that

$$\int_{-\infty}^{+\infty} \delta(x)\, dx = 1. \tag{A.5}$$

Unfortunately, this property of $\delta(x)$ is not compatible with the definition (A.4). In fact, from Eq. (A.4) it follows that the integral must be equal to zero. In other words, it does not exist a function $\delta(x)$ which satisfies both Eq. (A.4) and Eq. (A.5). Dirac suggested that a way to circumvent this problem is to interpret the integral of Eq. (A.5) as

$$\int_{-\infty}^{+\infty} \delta(x)\, dx = \lim_{\epsilon \to 0^+} \int_{-\infty}^{+\infty} \delta_\epsilon(x)\, dx, \tag{A.6}$$

where $\delta_\epsilon(x)$ is a generic function of both x and ϵ such that

$$\lim_{\epsilon \to 0^+} \delta_\epsilon(x) = \begin{cases} +\infty & \text{for } x = 0 \\ 0 & \text{for } x \neq 0 \end{cases}, \tag{A.7}$$

$$\int_{-\infty}^{+\infty} \delta_\epsilon(x)\, dx = 1. \tag{A.8}$$

Thus, the Dirac delta function $\delta(x)$ is a "generalized function" (but, strictly-speaking, not a function) which satisfy Eqs. (A.4) and (A.5) with the caveat that the integral in Eq. (A.5) must be interpreted according to Eq. (A.6) where the functions $\delta_\epsilon(x)$ satisfy Eqs. (A.7) and (A.8).

There are infinite functions $\delta_\epsilon(x)$ which satisfy Eqs. (A.7) and (A.8). Among them there is, for instance, the following Gaussian

$$\delta_\epsilon(x) = \frac{1}{\epsilon\sqrt{\pi}} e^{-x^2/\epsilon^2}, \tag{A.9}$$

which clearly satisfies Eq. (A.7) and whose integral is equal to 1 for any value of ϵ. Another example is the function

$$\delta_\epsilon(x) = \begin{cases} \frac{1}{\epsilon} & \text{for } |x| \leq \epsilon/2 \\ 0 & \text{for } |x| > \epsilon/2 \end{cases}, \tag{A.10}$$

which again satisfies Eq. (A.7) and whose integral is equal to 1 for any value of c. In the following we shall use Eq. (A.10) to study the properties of the Dirac delta function.

Dirac Function and the Integrals

According to the approach of Dirac, the integral involving $\delta(x)$ must be interpreted as the limit of the corresponding integral involving $\delta_\epsilon(x)$, namely

$$\int_{-\infty}^{+\infty} \delta(x) f(x) \, dx = \lim_{\epsilon \to 0^+} \int_{-\infty}^{+\infty} \delta_\epsilon(x) f(x) \, dx, \qquad (A.11)$$

for any function $f(x)$. It is then easy to prove that

$$\int_{-\infty}^{+\infty} \delta(x) f(x) \, dx = f(0). \qquad (A.12)$$

by using Eq. (A.10) and the mean value theorem. Similarly one finds

$$\int_{-\infty}^{+\infty} \delta(x - c) f(x) \, dx = f(c). \qquad (A.13)$$

Several other properties of the Dirac delta function $\delta(x)$ follow from its definition. In particular

$$\delta(-x) = \delta(x), \qquad (A.14)$$

$$\delta(a\,x) = \frac{1}{|a|} \delta(x) \quad \text{with } a \neq 0, \qquad (A.15)$$

$$\delta(f(x)) = \sum_i \frac{1}{|f'(x_i)|} \delta(x - x_i) \quad \text{with } f(x_i) = 0. \qquad (A.16)$$

Dirac Function in D Spatial Dimensions

Up to now we have considered the Dirac delta function $\delta(x)$ with only one variable x. It is not difficult to define a Dirac delta function $\delta^{(D)}(\mathbf{r})$ in the case of a D-dimensional domain $\mathbb{R}^D$, where $\mathbf{r} = (x_1, x_2, \ldots, x_D) \in \mathbb{R}^D$ is a D-dimensional vector:

$$\delta^{(D)}(\mathbf{r}) = \begin{cases} +\infty & \text{for } \mathbf{r} = \mathbf{0} \\ 0 & \text{for } \mathbf{r} \neq \mathbf{0} \end{cases} \qquad (A.17)$$

and

$$\int_{\mathbb{R}^D} \delta^{(D)}(\mathbf{r})\, d^D\mathbf{r} = 1. \tag{A.18}$$

Notice that sometimes $\delta^{(D)}(\mathbf{r})$ is written using the simpler notation $\delta(\mathbf{r})$. Clearly, also in this case one must interpret the integral of Eq. (A.18) as

$$\int_{\mathbb{R}^D} \delta^{(D)}(\mathbf{r})\, d^D\mathbf{r} = \lim_{\epsilon \to 0^+} \int_{\mathbb{R}^D} \delta_\epsilon^{(D)}(\mathbf{r})\, d^D\mathbf{r}, \tag{A.19}$$

where $\delta_\epsilon^{(D)}(\mathbf{r})$ is a generic function of both $\mathbf{r}$ and ϵ such that

$$\lim_{\epsilon \to 0^+} \delta_\epsilon^{(D)}(\mathbf{r}) = \begin{cases} +\infty & \text{for } \mathbf{r} = \mathbf{0} \\ 0 & \text{for } \mathbf{r} \neq \mathbf{0} \end{cases}, \tag{A.20}$$

$$\lim_{\epsilon \to 0^+} \int \delta_\epsilon^{(D)}(\mathbf{r})\, d^D\mathbf{r} = 1. \tag{A.21}$$

Several properties of $\delta(x)$ remain valid also for $\delta^{(D)}(\mathbf{r})$. Nevertheless, some properties of $\delta^{(D)}(\mathbf{r})$ depend on the space dimension D. For instance, one can prove the remarkable formula

$$\delta^{(D)}(\mathbf{r}) = \begin{cases} \frac{1}{2\pi} \nabla^2 (\ln |\mathbf{r}|) & \text{for } D = 2 \\ -\frac{1}{D(D-2)V_D} \nabla^2 \left(\frac{1}{|\mathbf{r}|^{D-2}} \right) & \text{for } D \geq 3 \end{cases}, \tag{A.22}$$

where $\nabla^2 = \frac{\partial^2}{\partial x_1^2} + \frac{\partial^2}{\partial x_2^2} + \cdots + \frac{\partial^2}{\partial x_D^2}$ and $V_D = \pi^{D/2}/\Gamma(1 + D/2)$ is the volume of a D-dimensional ipersphere of unitary radius, with $\Gamma(x)$ the Euler Gamma function. In the case $D = 3$ the previous formula becomes

$$\delta^{(3)}(\mathbf{r}) = -\frac{1}{4\pi} \nabla^2 \left(\frac{1}{|\mathbf{r}|} \right), \tag{A.23}$$

which can be used to transform the Gauss law of electromagnetism from its integral form to its differential form.

Appendix B
Complex Numbers

Set of Complex Numbers

Complex numbers were introduced in 1545 by Girolamo Cardano as an auxiliary tool for determining the real solutions of some third-degree algebraic equations. At that time, complex numbers were also used by others Italian mathematicians such as Scipione del Ferro, Raffele Bombelli, Niccolo Tartaglia, and Ludovico Ferrari, to determine the real solutions not only of third-degree algebric equations but also those of the fourth degree. As mentioned, complex numbers initially did not come considered as serious numbers but only as useful tool for solving equations. In the XVIII century Abraham de Moivre and Leonhard Euler began to provide complex numbers with a theoretical basis, until they assumed full citizenship in the mathematical world with the works of Carl Friedrich Gauss.

The set of complex numbers $\mathbb{C}$ is defined as

$$\mathbb{C} = \left\{ z : z = x + i\,y, \text{ where } x, y \in \mathbb{R} \text{ while } i = \sqrt{-1} \right\}. \tag{B.1}$$

It follows that a generic complex number z is given by

$$z = x + i\,y, \tag{B.2}$$

where x and y are real numbers while i is not a real number. The intrinsically complex number i, called imaginary unit, is defined as

$$i = \sqrt{-1}, \tag{B.3}$$

namely the square root of -1, which is obviously not a real number. Indeed, it does not exist a real number such that its square is equal to -1. The real numbers x and y of the complex number $z = x + iy$ are said respectively real part and imaginary part of the complex number z. Sometimes one writes $x = Re[z]$ and $y = Im[z]$.

© The Editor(s) (if applicable) and The Author(s), under exclusive license to Springer Nature Switzerland AG 2022
L. Salasnich, *Modern Physics*, UNITEXT for Physics,
https://doi.org/10.1007/978-3-030-93743-0

171

The set of real numbers $\mathbb{R}$ is a subset of the set of complex numbers $\mathbb{C}$, that is

$$\mathbb{R} \subset \mathbb{C}. \tag{B.4}$$

Indeed, if the imaginary part of the complex number is zero, the complex number is in fact a real number. Conversely, if the real part of the complex number is zero, the complex number is not a real number and is also called imaginary pure. For instance, $z = i\,3$ is a purely imaginary complex number, while $z = 2$ is a complex number but also a real number.

By definition, the algebraic properties that are used in complex numbers are the same of real numbers. Therefore, for example,

$$x + i\,y = x + y\,i = i\,y + x = y\,i + x. \tag{B.5}$$

It is very important to remember is that the square of the imaginary unit i is equal to -1, that is

$$i^2 = -1. \tag{B.6}$$

Obviously it follows that

$$i^3 = i^2\,i = (-1)\,i = -i, \tag{B.7}$$

and similarly

$$i^4 = i^2\,i^2 = (-1)\,(-1) = +1 = 1. \tag{B.8}$$

The complex conjugate of a complex number $z = x + iy$ is indicated with z^* (sometimes also with $\bar{z}$) and it is defined as follows

$$z^* = x - iy. \tag{B.9}$$

Hence the complex conjugate z^* has the same real part as z but an imaginary part with opposite sign. The modulus, or absolute value, of a complex number $z = x + iy$ is denoted by $|z|$ and it is defined as follows

$$|z| = \sqrt{x^2 + y^2}. \tag{B.10}$$

So the modulus $|z|$ is definitely a non-negative real number. For example, given the complex number $z = 2 - 3i$, its modulus reads

$$|z| = \sqrt{2^2 + (-3)^2} = \sqrt{4 + 9} = \sqrt{13}. \tag{B.11}$$

It is then quite easy to prove that for a generic complex number z the following equalities hold

$$|z|^2 = z^*z = zz^*. \tag{B.12}$$

Gauss Plane

As previously seen, the complex number $z = x + iy$ is characterized from two real numbers x and y which uniquely determine z. Therefore it is possible to introduce a Cartesian plane, called the Gauss plane or complex plane, such that on the horizontal axis of the abscissa, called real axis, we put $x = Re[z]$ while on the vertical axis of the ordinates, said imaginary axis, we set $y = Im[z]$. In this way any point P on the Gauss plane is characterized by two coordinates x and y, i.e.

$$P = (x, y), \tag{B.13}$$

where x and y are obviously the real part and the imaginary part of the complex number $z = x + iy$. Thus, any complex number z is in one-to-one correspondence with a point P of the Gauss plane. Formally it can be written

$$z = x + iy \leftrightarrow P = (x, y). \tag{B.14}$$

The geometric meaning of the modulus $|z|$ of a complex number z is therefore evident: $|z| = \sqrt{x^2 + y^2}$ represents the distance of the number z from the complex number 0, which is in the origin of the axes of the Gauss plane, that is $0 \leftrightarrow (0, 0)$.

Polar Representation

We have seen that the complex number $z = x + iy$ is in one-to-one correspondence with a point $P = (x, y)$ of the Gaussian plane. There is also a polar representation $[r, \phi]$, where

$$x = r \cos(\phi) \qquad y = r \sin(\phi) \tag{B.15}$$

with $r = \sqrt{x^2 + y^2}$ the radius (distance) and $\phi = artan(y/x)$ the angle shown in the figure.

The Cartesian representation of a complex number z is

$$z = x + iy. \tag{B.16}$$

Since we can write

$$x = r \cos(\phi) \tag{B.17}$$
$$y = r \sin(\phi) \tag{B.18}$$

it follows that

$$z = r \cos(\phi) + i\, r \sin(\phi) = r\, (\cos(\phi) + i \sin(\phi)) \tag{B.19}$$

is the polar representation of the complex number z. Evidently the modulus $|z|$ of z is precisely the radius r:

$$|z| = r = \sqrt{x^2 + y^2}. \tag{B.20}$$

Euler Formula

An extremely important result due to Leonard Euler is the formula

$$e^{i\phi} = \cos(\phi) + i\sin(\phi). \tag{B.21}$$

It follows that the polar representation of the complex number z can be written compactly as

$$z = r\,e^{i\phi}. \tag{B.22}$$

From Euler's formula (B.21) it follows that

$$e^{i\pi} = \cos(\pi) + i\sin(\pi) = -1$$

i.e.

$$e^{i\pi} + 1 = 0$$

which is considered the most beautiful formula in mathematics because it relates 5 fundamental objects of mathematics: 0, 1, e, π, and i.

Proof of the Euler Formula

To prove the Euler formula we must assume that we can to do a Taylor-MacLaurin series development on the function $e^{i\phi}$ around $\phi = 0$. That is

$$
\begin{aligned}
e^{i\phi} &= 1 + (i\phi) + \frac{1}{2!}(i\phi)^2 + \frac{1}{3!}(i\phi)^3 + \frac{1}{4!}(i\phi)^4 + \frac{1}{5!}(i\phi)^5 + \cdots \\
&= 1 + i\phi - \frac{1}{2!}\phi^2 - i\frac{1}{3!}\phi^3 + \frac{1}{4!}\phi^4 + i\frac{1}{5!}\phi^5 + \cdots \\
&= \left(1 - \frac{1}{2!}\phi^2\frac{1}{4!}\phi^4 + \cdots\right) + i\left(\phi - \frac{1}{3!}\phi^3 + \frac{1}{5!}\phi^5 + \cdots\right) \\
&= \cos(\phi) + i\sin(\phi),
\end{aligned}
\tag{B.23}
$$

remembering the Taylor-MacLaurin series of $\cos(\phi)$ and $\sin(\phi)$.

De Moivre Formula

The formula of de Moivre

$$(\cos(\phi) + i\sin(\phi))^n = \cos(n\phi) + i\sin(n\phi) \tag{B.24}$$

is a direct consequence of the Euler formula and of the fact of using the rules of algebra to complex numbers. In fact

$$(\cos(\phi) + i\sin(\phi))^n = \left(e^{i\phi}\right)^n = e^{in\phi} = \cos(n\phi) + i\sin(n\phi) \tag{B.25}$$

The de Moivre formula can be used to find complex solutions of simple algebraic equations of simple algebraic equations. For example, the equation

$$z^3 = 2 \tag{B.26}$$

can be solved by writing $z = re^{i\phi}$ so that

$$r^3 e^{3i\phi} = 2 e^{i2\pi n}, \tag{B.27}$$

with n an arbitrary integer. It follows that $r^3 = 2$ but also that $\phi = 2\pi n/3$. And so we get $r = 2^{1/3}$ and three values of ϕ given by

$$\phi_1 = 0, \qquad \phi_2 = -\frac{2\pi}{3}, \qquad \phi_3 = 2. \tag{B.28}$$

Fundamental Theorem of Algebra

The fundamental theorem of algebra states that the algebraic equation

$$a_n z^n + a_{n-1} z^{n-1} + \cdots + a_1 z + a_0 = 0, \tag{B.29}$$

with z unknown and coefficients $a_0, a_1, \ldots, a_{n-1}, a_n$ known, always admits n complex solutions.

For example, the equation

$$z^2 + 4 = 0$$

admits the two complex solutions $z_1 = -2i$ and $z_2 = 2i$.

Another example, the equation

$$z(z^2 + 3) = 0$$

admits the three complex solutions $z_1 = 0$, $z_2 = -i\sqrt{3}$ and $z_3 = i\sqrt{3}$.

Complex Functions

A function $\zeta = f(z)$ with complex domain $\mathbb{C}$ and complex codomain $\mathbb{C}$ can be formally written as

$$f : \mathbb{C} \to \mathbb{C}. \tag{B.30}$$

A function $\zeta = f(x, y, z)$ with real domain $\mathbb{R}^3$ and real codomain $\mathbb{R}$ can be formally written as

$$f : \mathbb{R}^3 \to \mathbb{R}. \tag{B.31}$$

As we shall see in the main text, the solutions of the independent Schrödinger equation, also called stationary Schrödinger equation, are exactly of this kind.

A function $\zeta = f(x, y, z, t)$ with real domain $\mathbb{R}^4$ and complex codomain $\mathbb{C}$ can be formally written as

$$f : \mathbb{R}^4 \to \mathbb{C}. \tag{B.32}$$

As we shall see, the solutions of the time-dependent Schrödinger equation are exactly of this kind.

Appendix C
Fourier Transform

Geometric and Taylor Series

It was known from the times of Archimedes that, in some cases, the infinite sum of decreasing numbers can produce a finite result. But it was only in 1593 that the mathematician Francois Viete gave the first example of a function, $f(x) = 1/(1 - x)$, written as the infinite sum of power functions. This function is nothing else than the geometric series, given by

$$\frac{1}{1-x} = \sum_{n=0}^{\infty} x^n, \quad \text{for } |x| < 1. \tag{C.1}$$

In 1714 Brook Taylor suggested that any real function $f(x)$ which is infinitely differentiable in x_0 and sufficiently regular can be written as a series of powers, i.e.

$$f(x) = \sum_{n=0}^{\infty} c_n (x - x_0)^n, \tag{C.2}$$

where the coefficients c_n are given by

$$c_n = \frac{1}{n!} f^{(n)}(x_0), \tag{C.3}$$

with $f^{(n)}(x)$ the n-th derivative of the function $f(x)$. The series (C.2) is now called Taylor series and becomes the so-called Maclaurin series if $x_0 = 0$. Clearly, the geometric series (C.1) is nothing else than the Maclaurin series, where $c_n = 1$. We observe that it is quite easy to prove the Taylor series: it is sufficient to suppose that Eq. (C.2) is valid and then to derive the coefficients c_n by calculating the derivatives of $f(x)$ at $x = x_0$; in this way one gets Eq. (C.3).

© The Editor(s) (if applicable) and The Author(s), under exclusive license to Springer Nature Switzerland AG 2022
L. Salasnich, *Modern Physics*, UNITEXT for Physics,
https://doi.org/10.1007/978-3-030-93743-0

Fourier Series

In 1807 Jean Baptiste Joseph Fourier, who was interested on wave propagation and periodic phenomena, found that any sufficiently regular real function function $f(x)$ which is periodic, i.e. such that

$$f(x + L) = f(x), \tag{C.4}$$

where L is the periodicity, can be written as the infinite sum of sinusoidal functions, namely

$$f(x) = \frac{a_0}{2} + \sum_{n=1}^{\infty} \left[a_n \cos\left(n\frac{2\pi}{L}x\right) + b_n \sin\left(n\frac{2\pi}{L}x\right) \right], \tag{C.5}$$

where

$$a_n = \frac{2}{L} \int_{-L/2}^{L/2} f(y) \cos\left(n\frac{2\pi}{L}y\right) dy, \tag{C.6}$$

$$b_n = \frac{2}{L} \int_{-L/2}^{L/2} f(y) \sin\left(n\frac{2\pi}{L}y\right) dy. \tag{C.7}$$

It is quite easy to prove also the series (C.5), which is now called Fourier series. In fact, it is sufficient to suppose that Eq. (C.5) is valid and then to derive the coefficients a_n and b_n by multiplying both side of Eq. (C.5) by $\cos\left(n\frac{2\pi}{L}x\right)$ and $\sin\left(n\frac{2\pi}{L}x\right)$ respectively and integrating over one period L; in this way one gets Eqs. (C.6) and (C.7).

It is important to stress that, in general, the real variable x of the function $f(x)$ can represent a space coordinate but also a time coordinate. In the former case L gives the spatial periodicity and $2\pi/L$ is the wavenumber, while in the latter case L is the time periodicity and $2\pi/L$ the angular frequency.

Complex Representation of the Fourier Series

Taking into account the Euler formula

$$e^{in\frac{2\pi}{L}x} = \cos\left(n\frac{2\pi}{L}x\right) + i \sin\left(n\frac{2\pi}{L}x\right) \tag{C.8}$$

with $i = \sqrt{-1}$ the imaginary unit, Fourier observed that his series (C.5) can be re-written in the very elegant form

$$f(x) = \sum_{n=-\infty}^{+\infty} f_n \, e^{in\frac{2\pi}{L}x}, \tag{C.9}$$

where

$$f_n = \frac{1}{L} \int_{-L/2}^{L/2} f(y) \, e^{-in\frac{2\pi}{L}y} \, dy \tag{C.10}$$

are complex coefficients, with $f_0 = a_0/2$, $f_n = (a_n - ib_n)/2$ if $n > 0$ and $f_n = (a_{-n} + ib_{-n})/2$ if $n < 0$, thus $f_n^* = f_{-n}$.

Fourier Integral

The complex representation (C.9) suggests that the function $f(x)$ can be periodic but complex, i.e. such that $f : \mathbb{R} \to \mathbb{C}$. Moreover, one can consider the limit $L \to +\infty$ of infinite periodicity, i.e. a function which is not periodic. In this limit Eq. (C.9) becomes the so-called Fourier integral (or Fourier anti-transform)

$$f(x) = \frac{1}{2\pi} \int_{-\infty}^{+\infty} \tilde{f}(k) \, e^{ikx} \, dk \tag{C.11}$$

with

$$\tilde{f}(k) = \int_{-\infty}^{\infty} f(y) \, e^{-iky} \, dy \tag{C.12}$$

the Fourier transform of $f(x)$. To prove Eqs. (C.11) and (C.12) we write Eq. (C.9) taking into account Eq. (C.10) and we find

$$f(x) = \sum_{n=-\infty}^{+\infty} \left(\frac{1}{L} \int_{-L/2}^{L/2} f(y) \, e^{-in\frac{2\pi}{L}y} \, dy \right) e^{in\frac{2\pi}{L}x}. \tag{C.13}$$

Setting

$$k_n = n\frac{2\pi}{L} \quad \text{and} \quad \Delta k = k_{n+1} - k_n = \frac{2\pi}{L} \tag{C.14}$$

the previous expression of $f(x)$ becomes

$$f(x) = \frac{1}{2\pi} \sum_{n=-\infty}^{+\infty} \left(\int_{-L/2}^{L/2} f(y) \, e^{-ik_n y} \, dy \right) e^{ik_n x} \, \Delta k. \tag{C.15}$$

In the limit $L \to +\infty$ one has $\Delta k \to dk$, $k_n \to k$ and consequently

$$f(x) = \frac{1}{2\pi} \int_{-\infty}^{+\infty} \left(\int_{-\infty}^{+\infty} f(y)\, e^{-iky}\, dy \right) e^{ikx}\, dk, \tag{C.16}$$

which gives exactly Eqs. (C.11) and (C.12). Note, however, that one gets the same result (C.16) if the Fourier integral and its Fourier transform are defined multiplying them respectively with a generic constant and its inverse. Thus, we have found that any sufficiently regular complex function $f(x)$ of real variable x which is globally integrable, i.e. such that

$$\int_{-\infty}^{+\infty} |f(x)|\, dx < +\infty, \tag{C.17}$$

can be considered as the (infinite) superposition of complex monocromatic waves e^{ikx}. The amplitude $\tilde{f}(k)$ of the monocromatic wave e^{ikx} is the Fourier transform of $f(x)$.

$f(x)$	$\mathcal{F}[f(x)](k)$
0	0
1	$2\pi\delta(k)$
$\delta(x)$	1
$\Theta(x)$	$\frac{1}{ik} + \pi\,\delta(k)$
$e^{ik_0 x}$	$2\pi\,\delta(k - k_0)$
$e^{-x^2/(2a^2)}$	$a\sqrt{2\pi}\,e^{-a^2 k^2/2}$
$e^{-a|x|}$	$\frac{2a}{a^2 + k^2}$
$sgn(x)$	$\frac{2}{ik}$
$\sin(k_0 x)$	$\frac{\pi}{i}\,[\delta(k - k_0) - \delta(k + k_0)]$
$\cos(k_0 x)$	$\pi\,[\delta(k - k_0) + \delta(k + k_0)]$

Table: Fourier transforms $\mathcal{F}[f(x)](k)$ of simple functions $f(x)$, where $\delta(x)$ is the Dirac delta function, $sgn(x)$ is the sign function, and $\Theta(x)$ is the Heaviside step function.

Properties of the Fourier Transform

The Fourier transform $\tilde{f}(k)$ of a function $f(x)$ is sometimes denoted as $\mathcal{F}[f(x)](k)$, namely

$$\tilde{f}(k) = \mathcal{F}[f(x)](k) = \int_{-\infty}^{\infty} f(x)\, e^{-ikx}\, dx. \tag{C.18}$$

The Fourier transform $\mathcal{F}[f(x)](k)$ has many interesting properties. For instance, due to the linearity of the integral the Fourier transform is clearly a linear map:

$$\mathcal{F}[a\, f(x) + b\, g(x)](k) = a\, \mathcal{F}[f(x)](k) + b\, \mathcal{F}[g(x)](k). \tag{C.19}$$

Moreover, one finds immediately that

$$\mathcal{F}[f(x-a)](k) = e^{-ika}\,\mathcal{F}[f(x)](k), \tag{C.20}$$

$$\mathcal{F}[e^{ik_0x}f(x)](k) = \mathcal{F}[f(x)](k-k_0). \tag{C.21}$$

$$\mathcal{F}[x\,f(x)](k) = i\,\tilde{f}'(k), \tag{C.22}$$

$$\mathcal{F}[f^{(n)}(x)](k) = (ik)^n\,\tilde{f}(k), \tag{C.23}$$

where $f^{(n)}(x)$ is the n-th derivative of $f(x)$ with respect to x.

In the Table we report the Fourier transforms $\mathcal{F}[f(x)](k)$ of some elementary functions $f(x)$, including the Dirac delta function $\delta(x)$ and the Heaviside step function $\Theta(x)$. We insert also the sign function $sgn(x)$ defined as: $sgn(x) = 1$ for $x > 0$ and $sgn(x) = -1$ for $x < 0$.

Fourier Transform and Uncertanty Theorem

The table of Fourier transforms clearly shows that the Fourier transform localizes functions which is delocalized, while it delocalizes functions which are localized. In fact, the Fourier transform of a constant is a Dirac delta function while the Fourier transform of a Dirac delta function is a constant. In general, it holds the following uncertainty theorem

$$\Delta x\,\Delta k \geq \frac{1}{2}, \tag{C.24}$$

where

$$(\Delta x)^2 = \int_{-\infty}^{\infty} x^2\,|f(x)|^2\,dx - \left(\int_{-\infty}^{\infty} x\,|f(x)|^2\,dx\right)^2 \tag{C.25}$$

and

$$(\Delta k)^2 = \int_{-\infty}^{\infty} k^2\,|\tilde{f}(k)|^2\,dk - \left(\int_{-\infty}^{\infty} k\,|\tilde{f}(k)|^2\,dk\right)^2 \tag{C.26}$$

are the spreads of the wavepackets respectively in the space x and in the dual space k. This theorem is nothing else than the uncertainty principle of quantum mechanics formulated by Werner Heisenberg in 1927, where x is the position and k is the wavenumber. Another interesting and intuitive relationship is the Parseval identity, given by

$$\int_{-\infty}^{+\infty} |f(x)|^2 dx = \int_{-\infty}^{+\infty} |\tilde{f}(k)|^2 dk. \tag{C.27}$$

Fourier Transform of Space-Time Functions

The Fourier transform is often used in electronics. In that field of research the signal of amplitude f depends on time t, i.e. $f = f(t)$. In this case the dual variable of time t is the frequency ω and the fourier integral is usually written as

$$f(t) = \frac{1}{2\pi} \int_{-\infty}^{+\infty} \tilde{f}(\omega) \, e^{-i\omega t} \, dk \tag{C.28}$$

with

$$\tilde{f}(\omega) = \mathcal{F}[f(t)](\omega) = \int_{-\infty}^{\infty} f(t) \, e^{i\omega t} \, dt \tag{C.29}$$

the Fourier transform of $f(t)$. Clearly, the function $f(t)$ can be seen as the Fourier anti-transform of $\tilde{f}(\omega)$, in symbols

$$f(t) = \mathcal{F}^{-1}[\tilde{f}(\omega)](t) = \mathcal{F}^{-1}[\mathcal{F}[f(t)](\omega)](t), \tag{C.30}$$

which obviously means that the composition $\mathcal{F}^{-1} \circ \mathcal{F}$ gives the identity.

More generally, if the signal f depends on the 3 spatial coordinates $\mathbf{r} = (x, y, z)$ and time t, i.e. $f = f(\mathbf{r}, t)$, one can introduce Fourier transforms from $\mathbf{r}$ to $\mathbf{k}$, from t to ω, or both. In this latter case one obviously obtains

$$f(\mathbf{r}, t) = \frac{1}{(2\pi)^4} \int_{\mathbb{R}^4} \tilde{f}(\mathbf{k}, \omega) \, e^{i(\mathbf{k}\cdot\mathbf{x}-\omega t)} \, d^3k \, d\omega \tag{C.31}$$

with

$$\tilde{f}(\mathbf{k}, \omega) = \mathcal{F}[f(\mathbf{r}, t)](\mathbf{k}, \omega) = \int_{\mathbb{R}^4} f(\mathbf{r}, t) \, e^{-i(\mathbf{k}\cdot\mathbf{r}-\omega t)} \, d^3\mathbf{r} \, dt. \tag{C.32}$$

Also in this general case the function $f(\mathbf{r}, t)$ can be seen as the Fourier anti-transform of $\tilde{f}(\mathbf{k}, \omega)$, in symbols

$$f(\mathbf{r}, t) = \mathcal{F}^{-1}[\tilde{f}(\mathbf{k}, \omega)](\mathbf{r}, t) = \mathcal{F}^{-1}[\mathcal{F}[f(\mathbf{r}, t)](\mathbf{k}, \omega)](\mathbf{r}, t). \tag{C.33}$$

Appendix D
Differential Equations

First-Order ODE

A typical first-order ordinary differential equation (ODE) is given by

$$a(x)\, f'(x) + b(x)\, f(x) = g[f(x), x], \tag{D.1}$$

where $f(x)$ is the unknown function, while $a(x)$, $b(x)$ and $g[f(x), x]$ are known functions. This is a first-order ODE because it appears only the first derivative $f'(x)$ of the unknown function $f(x)$. For example, the equation might be

$$f'(x) = 2\sin(x) f(x)^2 + 3, \tag{D.2}$$

where clearly in this case we have that $a(x) = 1$, $b(x) = 0$ and $g[f(x), x] = 2\sin(x) f(x)^2 + 3$.

The most general first-order ODE with constant coefficients reads

$$a\, f'(x) + b\, f(x) = c, \tag{D.3}$$

where $f(x)$ is the unknown function, while a, b and c are known coefficients. For example, the equation might be

$$-f'(x) + 3\, f(x) = \pi, \tag{D.4}$$

where clearly in this case we have that $a = -1$, $b = 3$ and $c = \pi$. It is important to note that the coefficients a, b and c could also be complex numbers. In this case the unknown function $f(x)$ has a complex codomain. If $c = 0$ the ODE is said homogeneous.

The homogeneous first-oder ODE with constant coefficients

$$a\, f'(x) + b\, f(x) = 0, \tag{D.5}$$

L. Salasnich, *Modern Physics*, UNITEXT for Physics, https://doi.org/10.1007/978-3-030-93743-0

admits the general solution

$$f(x) = A e^{\kappa x} \tag{D.6}$$

where κ is the solution of the algebraic equation

$$a\kappa + b = 0 \quad \text{that is} \quad \kappa = -\frac{b}{a}, \tag{D.7}$$

while the arbitrary constant A is determined by fixing an initial condition to the unknown function $f(x)$. This result is demonstrated by verifying that once we enter Eq. (D.6) into the left hand side of the equal of the Eq. (D.5) we find zero only if κ satisfies Eq. (D.7).

As an example, we consider the homogeneous first-order ODE with constant coefficients

$$f'(x) + 4 f(x) = 0 \tag{D.8}$$

with the initial condition $f(0) = 3$. This equation admits the general solution

$$f(x) = A e^{\kappa x}, \tag{D.9}$$

where κ is the solution of the algebraic equation

$$\kappa + 4 = 0 \quad \text{i.e.} \quad \kappa = -4, \tag{D.10}$$

and therefore

$$f(x) = A e^{-4x}. \tag{D.11}$$

The initial condition $f(0) = 3$ implies $f(0) = A = 3$. Ultimately, the solution of the ODE results in

$$f(x) = 3 e^{-4x}. \tag{D.12}$$

Separation of Variables

The homogeneous first-order ODE of the type

$$a(x) f'(x) = d(x) p[f(x)], \tag{D.13}$$

can be formally solved by the method of the separation of variables. Given that $f'(x) = \frac{df}{dx}$ the equation can be rewritten formally as

$$a(x)\frac{df}{dx} = d(x) p[f] \tag{D.14}$$

and then also

$$\frac{df}{p[f]} = \frac{d(x)}{a(x)} dx. \tag{D.15}$$

This equation that involves the differentials df and dx can be integrated, giving

$$\int_{f(x_0)}^{f(x)} \frac{df}{p[f]} = \int_{x_0}^{x} \frac{d(x)}{a(x)} dx, \tag{D.16}$$

where $f(x_0)$ is the initial condition of $f(x)$ at x_0. If we can calculate the two integrals, we get the solution $f(x)$ sought.

For example, let us consider the ODE

$$x f'(x) = x^3 f(x)^2, \tag{D.17}$$

with initial condition $f(1) = 2$. Setting $f'(x) = \frac{df}{dx}$ the equation can be rewritten as

$$\frac{df}{dx} = x^2 f^2 \tag{D.18}$$

and then

$$\frac{df}{f^2} = x^2 dx. \tag{D.19}$$

This equation between differentials can be integrated

$$\int_{2}^{f(x)} \frac{df}{f^2} = \int_{1}^{x} x^2 dx \tag{D.20}$$

and we then obtain

$$\left[-\frac{1}{f} \right]_{2}^{f(x)} = \left[\frac{x^3}{3} \right]_{1}^{x}. \tag{D.21}$$

It follows that

$$-\frac{1}{f(x)} + \frac{1}{2} = \frac{x^3}{3} - \frac{1}{3} \tag{D.22}$$

and finally

$$f(x) = \frac{1}{\frac{5}{6} - \frac{x^3}{3}}. \tag{D.23}$$

Second-Order ODE

A typical second-order ordinary differential equation (ODE) is given by

$$a(x) \, f''(x) + b(x) \, f'(x) + c(x) \, f(x) = d[f(x), x], \tag{D.24}$$

where $f(x)$ is the unknown function, while $a(x)$, $b(x)$, $c(x)$, and $d[f(x), x]$ are known functions. This is a second-order ODE because it appears the second derivative $f''(x)$ of the unknown function $f(x)$ but not higher derivatives. For example, the equation might be

$$x^2 \, f''(x) + 3 \, f'(x) = 2 \sin(x) f(x)^4, \tag{D.25}$$

where clearly in this case we have that $a(x) = x^2$, $b(x) = 3$, $c(x) = 0$, and $d[f(x), x] = 2 \sin(x) f(x)^4$.

The most general second-order ODE with constant coefficients reads

$$a \, f''(x) + b \, f'(x) + c \, f(x) = d, \tag{D.26}$$

where $f(x)$ is the unknown function, while a, b, c, and d are known coefficients. For example, the equation might be

$$3 \, f''(x) - f'(x) + 2 \, f(x) = -7, \tag{D.27}$$

where clearly in this case we have that $a = 3$, $b = -1$, $c = 2$, and $d = -7$. It is important to note that the coefficients a, b, c, d could also be complex numbers. In this case also the unknown function $f(x)$ will be a function with complex codomain. If $d = 0$ the ODE is called homogeneous.

The homogeneous second-order ODE with constant coefficients

$$a \, f'(x) + b \, f'(x) + c \, f(x) = 0, \tag{D.28}$$

admits the general solution

$$f(x) = A e^{\kappa_1 x} + B e^{\kappa_2 x} \tag{D.29}$$

where κ_1 and κ_2 are the two complex solutions of the algebraic equation

$$a\kappa^2 + b\kappa + c = 0, \tag{D.30}$$

while the arbitrary constants A and B are determined by fixing two initial conditions to the unknown function $f(x)$. This result is demonstrated by verifying that once inserted Eq. (D.29) into the left-hand side of the Eq. (D.28) zero is found only if κ satisfies Eq. (D.30).

As an example, we consider the following homogeneous second-order ODE with constant coefficients

$$f''(x) + 4 \, f(x) = 0 \tag{D.31}$$

with initial conditions $f(0) = 1$ and $f'(0) = 0$. This equation admits the general solution

$$f(x) = Ae^{\kappa_1 x} + Be^{\kappa_2 x} \tag{D.32}$$

where κ_1 and κ_2 are the two complex solutions of the algebraic equation

$$\kappa^2 + 4 = 0, \tag{D.33}$$

i.e. $\kappa_1 = -2i$ and $\kappa_2 = 2i$, and therefore

$$f(x) = Ae^{-i2x} + Be^{i2x}. \tag{D.34}$$

In this example we also have

$$f'(x) = -i2\, Ae^{-i2x} + i2Be^{i2x}. \tag{D.35}$$

The initial condition $f'(0) = 0$ implies

$$f'(0) = -i2\, A + i2B = 0 \tag{D.36}$$

and therefore

$$A = B. \tag{D.37}$$

Furthermore, the initial condition $f(0) = 1$ implies.

$$f(0) = A + B = 2\, A = 1, \tag{D.38}$$

and thus

$$A = B = \frac{1}{2}. \tag{D.39}$$

In conclusion, the solution of the ODE reads

$$f(x) = \frac{1}{2}\left(e^{-i2x} + e^{i2x}\right) = \cos(2x). \tag{D.40}$$

Newton Law as a Second-Order ODE

It is important to stress that the familiar Newton law

$$\mathbf{F} = m\,\mathbf{a} \tag{D.41}$$

is to all intents and purposes a second-order ODE. In fact it can be written in full as

$$\mathbf{F}\left(\mathbf{r}(t), \frac{d\mathbf{r}(t)}{dt}\right) = m\frac{d^2\mathbf{r}(t)}{dt^2} \tag{D.42}$$

where the unknown function is the radius vector $\mathbf{r}(t)$ as a function of time t.

In fact, the acceleration appears in Newton's law $\mathbf{a}(t)$ at time t. This can be written as the second derivative with respect to time t of the position vector $\mathbf{r}(t)$, i.e.

$$\mathbf{a}(t) = \frac{d^2\mathbf{r}(t)}{dt^2}. \tag{D.43}$$

Moreover, in certain cases, the force $\mathbf{F}$ can depend not only on the position $\mathbf{r}(t)$ but also on the velocity $\mathbf{v}(t)$, and therefore on the first derivative of the position vector, since $\mathbf{v}(t) = d\mathbf{r}(t)/dt$.

Partial Differential Equations

A partial differential equation (PDE) is an equation that involves the partial derivatives of an unknown function of several independent variables. For instance, in the differential equation

$$\frac{\partial f(x, y)}{\partial x} + \frac{\partial f(x, y)}{\partial y} = 2x^2 y \tag{D.44}$$

the partial derivatives of the unknown function $f(x, y)$ appear. This PDE is of the first order because only the first partial derivatives appear.

In general, a PDE is of order n if in the equation it appears the n-th partial derivative with respect to some variable. Usually in Physics the unknown function depends on the three spatial coordinates x, y, z and the time t. The unknown function can be a scalar quantity, that is of the type $f(x, y, z, t)$ or a vector quantity, that is of the type $\mathbf{v}(x, y, z, t)$. Compactly, we can rewrite the two functions as $f(\mathbf{r}, t)$ and $\mathbf{v}(\mathbf{r}, t)$, respectively. In Physics, a quantity that depends on the three spatial coordinates is called a field. Thus, one can have scalar fields and vector fields. More generally, we can also have tensor fields.

Wave Equation

A typical example of EDP from Physics is the wave equation, also called d'Alambert equation, given by

$$\left(\frac{1}{c^2}\frac{\partial^2}{\partial t^2} - \nabla^2\right) f(\mathbf{r}, t) = 0, \tag{D.45}$$

where

$$\nabla^2 = \frac{\partial^2}{\partial x^2} + \frac{\partial^2}{\partial y^2} + \frac{\partial^2}{\partial z^2} \tag{D.46}$$

is known as Laplace differential operator, or Laplacian, sometimes also called the nabla. The constant c is the velocity of propagation of the wave. In the case of electromagnetic waves c is the speed of light. In the case of sound waves, c is the speed of sound. This equation can be written in an even more compact way

$$\Box f(\mathbf{r}, t) = 0, \tag{D.47}$$

where

$$\Box = \frac{1}{c^2} \frac{\partial^2}{\partial t^2} - \nabla^2 \tag{D.48}$$

is known as d'Alambert differential operator, or d'Alambertian. Equation (D.47) admits the following complex solution, called monochromatic plane wave

$$f(\mathbf{r}, t) = f_0 \, e^{i(\mathbf{k} \cdot \mathbf{r} - \omega t)}, \tag{D.49}$$

where f_0 is an arbitrary constant, called the amplitude of the wave, $\mathbf{k}$ is called the wave vector, and ω is called the angular frequency of the wave. The wave vector and the angular frequency are not independent. In fact between them there is the following relationship

$$\omega = c k \tag{D.50}$$

called the dispersion relation, with $k = |\mathbf{k}|$. The dispersion relation is obtained immediately by inserting the function (D.49) in Eq. (D.47). As we have already seen, by setting $\omega = 2\pi\nu$ and $k = 2\pi/\lambda$ the relation can be rewritten as

$$\lambda \nu = c. \tag{D.51}$$

Since the d'Alambertian operator is linear, taking into account Euler formula of complex numbers, it is immediate to verify that also the functions

$$f(\mathbf{r}, t) = A \cos(\mathbf{k} \cdot \mathbf{r} - \omega t) \tag{D.52}$$

and

$$f(\mathbf{r}, t) = B \sin(\mathbf{k} \cdot \mathbf{r} - \omega t) \tag{D.53}$$

are solutions of the wave equation. Also in this case the dispersion relation $\omega = ck$ holds, and A and B are arbitrary constants. More generally, it follows immediately that also

$$f(\mathbf{r}, t) = A \cos(\mathbf{k} \cdot \mathbf{r} - \omega t) + B \sin(\mathbf{k} \cdot \mathbf{r} - \omega t) \tag{D.54}$$

is solution, but also

$$f(\mathbf{r}, t) = f_0 \, e^{i(\mathbf{k} \cdot \mathbf{r} - \omega t)} + f_1 \, e^{-i(\mathbf{k} \cdot \mathbf{r} - \omega t)} \tag{D.55}$$

is a solution of (D.47) with f_0 and f_1 arbitrary constants.

Diffusion Equation

Another typical example of PDE is the diffusion equation, also called heat equation, given by

$$\frac{\partial}{\partial t} f(\mathbf{r}, t) = D \nabla^2 f(\mathbf{r}, t), \tag{D.56}$$

where D is a constant, usually real, that is the so-called diffusion coefficient. This equation resembles the equation of Schrödinger equation of a quantum particle in the absence of an external potential. In the case of the Schrödinger equation the coefficient D is a complex number complex number given by

$$D = \frac{i\hbar}{2m}, \tag{D.57}$$

where i is the imaginary unit, $\hbar$ is the reduced Plank constant, and m is the mass of the quantum particle.

Considering the case of a single spatial coordinate, the diffusion eqution becomes

$$\frac{\partial}{\partial t} f(x, t) = D \frac{\partial^2}{\partial x^2} f(x, t). \tag{D.58}$$

Assuming that at time $t = 0$ the initial condition is Gaussian

$$f(x, t = 0) = e^{-x^2}, \tag{D.59}$$

it can be verified that the solution at time t is

$$f(x, t) = \frac{1}{\sqrt{\zeta(t)}} e^{-\frac{x^2}{\zeta(t)}}, \tag{D.60}$$

where

$$\zeta(t) = 1 + 2Dt. \tag{D.61}$$

So if $D > 0$ the solution widens and lowers as time increases, that is, it diffuses.

References

1. Bjorken, J.D., Drell, S.D.: Relativistic Quantum Mechanics. McGraw-Hill (1964)
2. Blackett, P.M.S., Occhialini, G.P.S.: Proc. R. Soc. Lond. A **139**, 699 (1933)
3. Boltzmann, L.: Sitzungsberichte der Kaiserlichen Akademie der Wissenschaften in Wien. Mathematisch-Naturwissenschaftliche Classe **66**, 275 (1872)
4. Born, M.: Z. Phys. **37**, 863 (1926)
5. Born, M.: Z. Phys. **38**, 803 (1926)
6. Born, M., Jordan, P.: Z. Phys. **34**, 858 (1925)
7. Born, M., Heisenberg, W., Jordan, P.: Z. Phys. **35**, 557 (1926)
8. Bohr, N.: Phil. Mag. **26**, 1 (1913)
9. Bransden, B.H., Joachain, C.J.: Physics of Atoms and Molecules. Prentice Hall (2003)
10. Cohen-Tannoudji, C., Dui, B., Laloe, F.: Quantum Machanics. Wiley (2019)
11. Compton, A.H.: Phys. Rev. **21**, 483 (1923)
12. Davisson, C.J., Germer, L.H.: Proc. Natl. Acad. Sci. **14**, 317 (1928)
13. Dirac, P.A.M.: Proc. R. Soc. Lond. A **114**, 243 (1927)
14. Dirac, P.A.M.: The Principles of Quantum Mechanics. First published in 1928. Oxford University Press (1982)
15. De Broglie, L.: Ann. Phys. **10**, 22 (1925)
16. Einstein, A.: Ann. Phys. **17**, 891 (1905)
17. Einstein, A.: Ann. Phys. **17**, 132 (1905)
18. Einstein, A.: Sitzungsberichte der Preussischen Akademie der Wissenschaften zu Berlin, Part 2, 844 (1915)
19. Einstein, A.: Verh. Dtsch. Phys. Ges. **18**, 318 (1916)
20. Fitzgerald, G.F.: Science **13**, 390 (1889)
21. Gibbs, J.W.: Elementary Principles in Statistical Mechanics (Charles Scribner's Sons, 1902). Cambridge University, Press (2010)
22. Heisenberg, W.: Z. Phys. **33**, 879 (1925)
23. Hilbert, D.: Nachrichten von der Gesellschaft der Wissenschaften zu Gottingen. Mathematisch-Physikalische Klasse **3**, 395 (1915)
24. Huang, K.: Statistical Mechanics. Wiley (1987)
25. Landau, L.D., Lifshitz, E.M.: The Classical Theory of Fields. Pergamon (1980)
26. L.D. Landau and E.M. Lifshitz, Quantum Mechanics: Non-Relativistic Theory (Pergamon, 1981)
27. Landau, L.D., Lifshitz, E.M.: Statistical Physics. Pergamon (1980)

© The Editor(s) (if applicable) and The Author(s), under exclusive license
to Springer Nature Switzerland AG 2022
L. Salasnich, *Modern Physics*, UNITEXT for Physics,
https://doi.org/10.1007/978-3-030-93743-0

28. Lorentz, H.A.: Proc. R. Neth. Acad. Arts Sci. **6**, 809 (1904)
29. Maxwell, J.C.: Philos. Trans. R. Soc. Lond. **157**, 49 (1867)
30. McGervey, J.D.: Introduction to Modern Physics. Academic Press (1983)
31. Merli, P.G., Missiroli, G.F., Pozzi, G.: Am. J. Phys. **44**, 306 (1976)
32. Michelson, A.A., Morley, E.W.: Am. J. Sci. **34**, 333 (1887)
33. Michelson, A.A., Morley, E.W.: Am. J. Sci. **34**, 427 (1887)
34. Planck, M.: Verh. Dtsch. Phys. Ges. **2**, 237 (1900)
35. Poincare, H.: Revue de Metaphysique et de Morale **6**, 1 (1898)
36. Rindler, W.: Introduction to Special Relativity. Oxford University Press (1991)
37. Robinett, R.W.: Quantum Mechanics: Classical Results, Modern Systems, and Visualized Examples. Oxford University Press (2006)
38. Robnik, M., Salasnich, L.: J. Phys. A **30**, 1711 (1997)
39. Robnik, M., Salasnich, L.: J. Phys. A **30**, 1719 (1997)
40. Rydberg, J.R.: Proc. R. Swed. Acad. Sci. **23**, 1 (1889)
41. Ryder, L.: Introduction to General Relativity. Cambridge University Press (2009)
42. Sakurai, J.J., Napolitano, J.: Modern Quantum Mechanics. Cambridge University Press (2020)
43. Salasnich, L.: Quantum Physics of Light and Matter: Photons, Atoms, and Strongly Correlated Systems. Springer (2017)
44. Schrödinger, E.: Phys. Rev. **28**, 1049 (1926)
45. Serway, R.A., Moses, C.J., Moyer, C.A.: Modern Physics. Brooks/Cole Publishing Company (2004)
46. Tonomura, A., Endo, J., Matsuda, T., Kawasaki, T., Ezawa, H.: Am. J. Phys. **57**, 117 (1989)
47. von Neumann, J.: Mathematical Foundations of Quantum Mechanics. First published in 1932. Princeton University Press (2018)

Index

© The Editor(s) (if applicable) and The Author(s), under exclusive license
to Springer Nature Switzerland AG 2022
L. Salasnich, *Modern Physics*, UNITEXT for Physics,
https://doi.org/10.1007/978-3-030-93743-0

Printed in the United States
by Baker & Taylor Publisher Services